Karl Cammann

Working with Ion-Selective Electrodes

Chemical Laboratory Practice

Translated from the German
by Albert H. Schroeder

With 65 Figures

Springer-Verlag
Berlin Heidelberg New York 1979

Karl Cammann
Abteilung für Analytische Chemie, Universität Ulm
D-7900 Ulm, Germany

Albert H. Schroeder
IBM, T. J. Watson Research Center
Yorktown Heights, NY 10598, USA

Enlarged and revised translation of the second edition of
Das Arbeiten mit ionenselektiven Elektroden by K. Cammann
in the series *Anleitungen für die chemische Laboratoriumspraxis*,
Vol. 13 (Springer-Verlag 1977)

Library of Congress Cataloging in Publication Data. Cammann, Karl. Working with ion-selective electrodes. Translation of the 2nd rev. and enl. ed. of Das Arbeiten mit ionenselektiven Elektroden. Includes bibliographical references and index. 1. Electrochemical analysis. 2. Electrodes, Ion selective. I. Title. QD115.C3413. 543'.087. 79-4335.

ISBN-13: 978-3-642-67278-1 **e-ISBN-13: 978-3-642-67276-7**
DOI: 10.1007/ 978-3-642-67276-7

Softcover reprint of the hardcover 1st edition 1979

Typesetting: J. Beltz, Hemsbach. Binding: J. Schäffer OHG, Grünstadt
2152/3140-543210

Preface to the Second German Edition

The relatively short time period in which the first edition was sold out demonstrates the interest which has developed in the electrochemical method of analysis. This edition differs from the first in its expanded treatment of the fields of gas sensors, enzyme electrodes and industrial applications. The extent to which this topic has grown in importance since 1973 is reflected in the 447 publications cited here, as compared to 162 references in the first edition. The nomenclature used here follows the recommendations of the IUPAC.

All firms mentioned in the text are sincerely thanked for their permission to use information and photographic materials. I wish to thank in particular Dr. R.P. Buck, University of North Carolina at Chapel Hill for many stimulating discussions, as well as Dr. G.A. Rechnitz, State University of New York at Buffalo and Dr. S. Mazur, University of Chicago for the opportunity to carry out a few fundamental measurements in their laboratories. For the preparation of the subject index I thank Ms. S. Schultz, M. A., University of Chicago.

Munich, January 1977

Preface to the First German Edition

Ion-selective electrodes allow the specific and quantitative determination of an almost unimaginable number of substances, ranging from simple inorganic ions through amino acids to complex organic molecules. The extent to which these ion-selective electrodes correspond to every analyst's ideal and dream is explored here. The wide range of applications, low material requirements and simplicity of the analytical procedure have not only brought ion-selective electrodes into the limelight of analytical chemistry, but have promoted their use as tools for physiologists, medical researchers, biologists, geologists, environmental protection specialists, etc. This volume should be of particular help to such people.

This book will also be of use to all students who have occasion to carry out chemical analyses during their educational careers. From a practical standpoint, many simplified analytical procedures are possible. In addition, the student can learn about a field of modern analytical chemistry not always discussed in detail in the textbooks. Those who have already had some experience with these new sensors will find references to new techniques.

This volume is not, and is not intended to be, a "cook book" for specific analyses using ion-selective electrodes for two reasons: The corresponding, detailed procedures are already available from the individual electrode manufacturers; in addition, such a work would only help a relatively small group of people who happened to be interested in the determination of the same ion in the same matrix. The reader will, however, find basic examples and suggestions throughout.

The main purpose of this volume is to acquaint the large circle of people who are interested in this method of analysis with its most important *fundamentals* and to offer *practical advice* for the optimum solution of the various problems which may arise. A brief introduction to the theoretical principles cannot be avoided if the reader is to be equipped with the understanding necessary for future development of these electrodes.

With this fundamental knowledge and a few additional tips concerning practical "know-how", the innovative experimentalist should be in a position to produce functional electrodes himself. This is especially important in physiological applications, since to date neither suitable microelectrodes nor sensors for the important amino acids are commercially available.

During his many years of activity in the applications department of an instrument manufacturer, the author came to realize that most problems which arise are not on account of the instruments themselves, but rather can be traced back to a lack of practical know-how on the part of the user.

3.2.1.3 Handling 55
3.2.1.4 Sample Preparation 57
3.2.2 Homogeneous Solid-State Membrane Electrodes For Ag^+, Cd^{2+}, Cu^{2+}, Pb^{2+}, S^{2-}, F^-, Cl^-, Br^-, I^-, SCN^-, CN^- Ions 58
3.2.2.1 Principles 58
3.2.2.2 Construction 59
3.2.2.3 Preparation of Solid-State Membrane Electrodes Based on Ag_2S 61
3.2.2.4 Characteristics 61
3.2.2.5 Handling 68
3.2.2.6 Sample Preparation 69
3.2.3 Heterogeneous Solid-State Membrane Electrodes For Ag^+, Cl^-, Br^-, I^-, CN^-, SCN^-, S^{2-} Ions 74
3.2.3.1 Construction 74
3.2.3.2 Preparation 75
3.2.3.3 Characteristics 75
3.2.3.4 Handling 77
3.3 Porous and Nonporous Supported Ion Exchanger and Neutral Carrier Membrane Electrodes 77
3.3.1 Ion-Exchangers For Ca^{2+}, Me^{2+} Cations and Cl^-, ClO_4^-, NO_3^-, BF_4^- Anions . . . 77
3.3.2 Neutral Carrier Compounds For Li^+, Na^+, K^+, Rb^+, Cs^+, NH_4^+, NR_4^+, Ca^{2+} and Ba^{2+} Cations . 79
3.3.3 Construction 80
3.3.4 Preparation of PVC Membrane Electrodes 82
3.3.5 Characteristics 82
3.3.6 Handling 87
3.3.7 Further Electro-Active Phases 87
3.4 Solid State Electrodes with Electroactive Coatings 89
3.5 Gas Sensors For CO_2, NH_3, SO_2, NO_2, HF, H_2S, HCN, etc. . . 93
3.5.1 Principles 93
3.5.2 Construction 94
3.5.3 Characteristics of Gas-Sensitive Electrodes 95
3.5.4 Handling 97
3.5.5 Sample Preparation 97
3.6 Bio-Sensors 98
3.6.1 Principles 98
3.6.2 Construction 99
3.6.3 Preparation of Enzyme Electrodes 99
3.6.4 Characteristics of Bio-Sensors 102
3.6.5 Sample Preparation with Bio-Sensors 103

4 Measuring Techniques with Ion-Selective Electrodes 105
4.1 Equivalent Circuit for a Cell with Liquid Junction 105
4.2 Measuring the EMF of an Electrochemical Cell 108
4.3 Choosing an EMF Measuring Device 110
4.4 Characteristics of Electrometer Amplifiers 114

4.4.1 Resolving Power 114
4.4.2 Insulation Problems 116
4.4.3 Charging Phenomena 116
4.4.4 Ground Loops 119

5 Analysis Techniques Using Ion-Selective Electrodes 121
5.1 Calibration Curves 122
5.1.1 Determination of Activity Using an Activity Calibration Curve . 122
5.1.2 Determination of Concentration Using a Concentration Calibration Curve 123
5.2 Direct Indication on the pH or pIon Scale of an Instrument . . 125
5.3 Titration Procedures for Determining Concentrations 127
5.3.1 Prerequisites 127
5.3.2 Titration Errors 128
5.3.3 Sample Preparation for Titrations 131
5.3.4 Titration to a Pre-determined EMF Value 132
5.3.4.1 On the Basis of a Titration Curve 132
5.3.4.2 On the Basis of a Concentration Cell Set-Up 133
5.3.5 "Chemically Linearized" Titration Curves (One Point Titration) 136
5.4 Concentration Determinations with the Help of a Standard Addition with a Known Electrode Slope *S* 137
5.4.1 Measuring the Change in EMF upon Addition of a Standard Solution to the Sample Solution 137
5.4.2 Measuring the Change in EMF upon Addition of the Sample Solution to a Standard Solution 140
5.5 Concentration Determinations with the Help of a Standard Addition with an Unknown Electrode Slope *S* 142
5.5.1 Method of Double Standard Addition 143
5.5.2 Method of Standard Addition with Subsequent Dilution . . . 144
5.6 Practical Example of the Addition Method: Sodium and Potassium Determination in Blood Serum 145
5.7 Concentration Determinations with the Help of a "Mathematically Linearized" Titration Curve 147
5.8 Practical Example of the Gran Extrapolation Method: Determination of Chloride in the ppm Range 155
5.8.1 Principles 156
5.8.2 Preparatory Work 156
5.8.3 Blank Determination 157
5.8.4 Chloride Content Determination of the Sample Solution . . . 157
5.9. Determination of Some Characteristic Electrode Parameters . 158
5.9.1 Determination of the Detection Limit 158
5.9.2 Determination of the Selectivity Coefficient 159

6 Applications of Ion-Selective Electrodes 162
6.1 Physiology, Biology, Medicine 162
6.1.1 Measurements in Extracellular Fluids 171

6.1.1.1 *In-vitro* Measurements 171
6.1.1.1.1 Sample Preparation 172
6.1.1.1.2 The Indicating Electrode 173
6.1.1.1.3 The Reference Electrode 173
6.1.1.2 *In-vivo* Measurements 174
6.1.2 Measurements of Intracellular Ion Activities 174
6.1.2.1 Preparation of Ion-Selective Microelectrodes 175
6.1.2.2 Guarded Signal Wires for Extremely High Ohmic Electrodes . . 176
6.1.2.3 Reference Microelectrodes 177
6.1.2.4 Special Amplifiers for Microelectrode Applications 178
6.2 Continuous Measurements in Industry and Environmental Research 179
6.2.1 Flow-thru Cells 180
6.2.1.1 Influence of Grounding 182
6.2.1.2 Influence of Temperature 183
6.2.1.3 Analysis Techniques with Flow-thru Measurements 186
6.2.1.4 Flow-thru Measurements without Reference Electrodes . . . 186
6.2.1.4.1 Direct Potentiometry 187
6.2.1.4.2 Concentration Determination through Standard Solution Addition 187
6.2.1.4.3 Indirect Concentration Determination 188
6.2.1.4.4 The Cyanide Monitor as an Example of Industrial On-Line Measurements 188
6.2.2 Concentration Determination Via a Continuous Titration . . 191

Outlook 194

Appendix 197
A.1 Concentration, Activity and Activity Coefficient 197
A.1.1 Preparing Accurate Activity Calibration Solutions 199
A.2 Survey of the Temperature Dependence of Commonly Used Reference Electrodes 203
A.3 Table for Evaluation Using Analysis Technique 5.4.1 204
A.4 Table for Evaluation Using Analysis Technique 5.4.2 206
A.5 Table for Evaluation Using Analysis Technique 5.5.1 207
A.6 Table for Evaluation Using Analysis Technique 5.5.2 208
A.7 Evaluation Table for Standard Addition + 1:1 Dilution . . . 209
A.8 Ion-Selective Electrode Manufacturers and/or Dealers 210

Literature 211
Subject Index 221
Index of Symbols Used 226

Introduction

Ion-selective electrodes are electrochemical half-cells in which a potential difference which is dependent on the concentration (more precisely the activity) of a particular ion in solution arises across the electrode/electrolyte interface. Their construction resembles that of the well-known pH glass electrode. Figure 1 shows a simple measuring set-up. Measurements are made in the same manner as pH measurements, requiring only a precise pH-meter and a suitable reference electrode. In place of the pH scale ($pH \equiv -\log a_{H^+}$) one can define ion-scales such as pNa^+, pCa^{2+}, pCu^{2+}, $pSCN^-$, etc. and call the corresponding electrodes pNa, pCa, etc. electrodes.

It is impossible to enumerate the many ions which can be determined with currently commercially available ion-selective electrodes because their number is continually increasing. Those cations or anions which are specifically and directly detected with these electrodes number about 30. In addition, there are those ions which, with the help of these 30, can be determined indirectly through chemical complexation, precipitation or biochemical reactions which serve to change the activity of the indicated ion in the solution equilibrium. Modified electrodes also make it possible to directly indicate gases such as NH_3, CO_2, NO_X, SO_2, etc., as well as organic compounds such as amino acids, amygdalin, cholesterol, urea, penicillin, etc. A current summary (Fig. 2) shows the periodic table with shaded boxes for elements which are directly indicated and boxes divided in half for elements which at present can be indicated only indirectly. The second row lists either the form in which the ion is specifically indicated or the chemical reaction with which the indirectly determined ion can be measured. No one has yet attempted to detect the remaining elements with the help of ion-selective electrodes as far as the author is aware (1973).

The possibility of directly and selectively detecting ions differing by their valence status (NH_4^+ and NO_3^-, S^{2-} and SO_4^{2-}, Cl^- and ClO_4^-) opens new horizons for analysis; the elements on the right half of the periodic table inclined toward anionic formations are precisely those which have most obstinately set back the progress of instrumental analysis. As *specific anion-detectors* ion-selective electrodes are almost without rival, considering the cost of all the necessary equipment (~$500) and the precision attainable when used for end-point determinations (within $< 1\%$). Ion-selective electrodes are especially interesting for environmental protection, since they provide a means of establishing the binding properties of the various measured ions.

Above all they are unchallenged for physiological measurements in biological cells dealing with the Na, K and Ca distributions so important in the intercellular transmission of information. Such measurements are difficult to obtain with other techniques due to physiological feed-back reactions induced by concentration changes during the sampling process. Also in this connection, microelectrodes for monitoring biologically

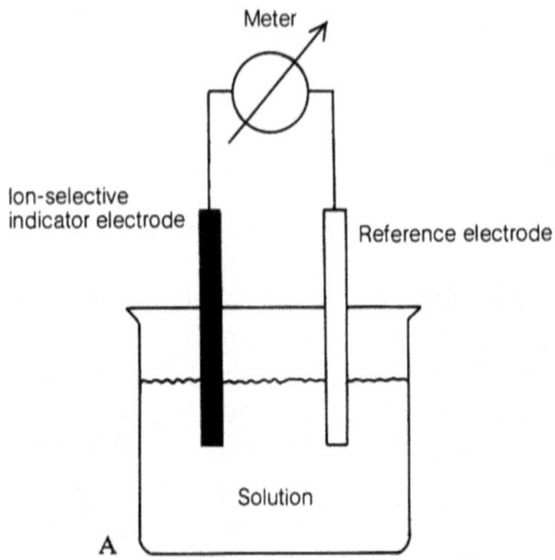

A

B

Fig. 1. A: Schematic diagram of a measuring set-up. **B:** Ion activity meter with all utensils needed for a quantitative determination. (Depending on the calibration, the indicated ion content of the sample solution can be read off of the logarithmically divided scale directly in M, meq/l, g/l, ppm, etc.)

important species (e.g. amino acids) appear to be an extremely valuable tool for researchers in the field of medicine and biology. The fact that direct measurement with these electrodes yields not the concentration but the activity of the appropriate ion in the sample solution need not be a drawback, since in recent years it has become apparent that even for physiological processes it is the difference in activity which is the driving force. Secondly, using known concentrations this technique provides the means to easily determine the activity coefficients, which again opens up a broad new field of application in physico-chemical basic research.

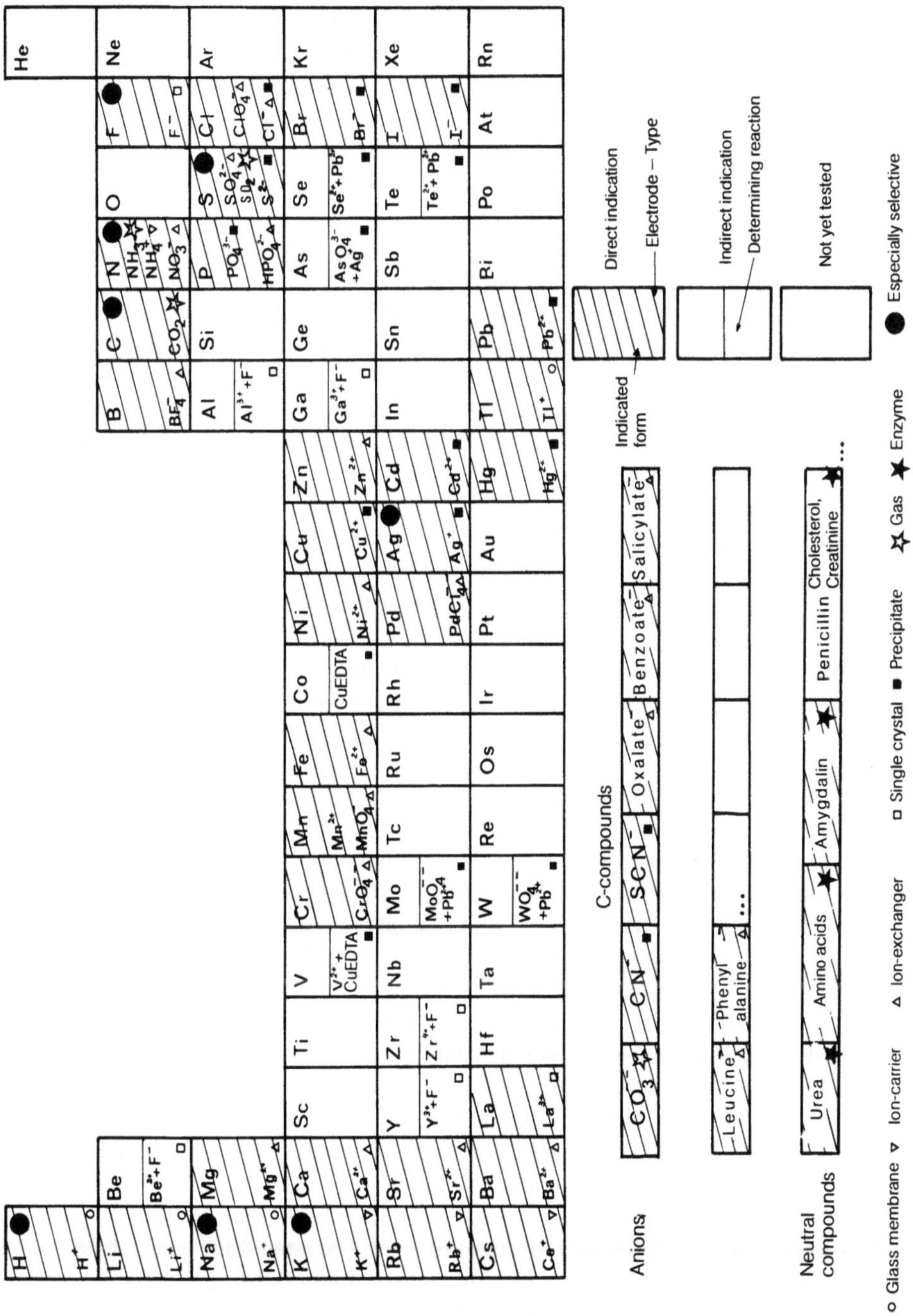

Fig. 2. Survey of the ions and selected neutral compounds which can presently be determined with the help of ion-selective electrodes

In many instances time-consuming sample preparations such as filtrations, distillations or extractions can be avoided by employing ion-selective electrodes. From an instrumental standpoint, no expensive, sophisticated instrumentation, i.e. integrators, compensators for non-linear calibration curves, etc. are needed. After the appropriate calibration, the concentration can be read directly from a voltmeter with a high impedance input.

EMF[1] measurements with an ion-selective electrode require very little external power for the needed electrometer amplifier. Another advantage is the easily attainable miniaturization of the measuring electrode and amplifier. The latter can be battery-operated, making field measurements quite easy. This attribute promotes their use as detectors for environmental protection. The relative accuracy, within 1 to 10% depending on the interferences present, in most cases completely suffices for trend analyses in the field (i.e. investigations of land, air and plants). In other cases they can render valuable service by allowing fast preliminary analyses far easier than any other analytical technique.

The sensitivity of ion-selective electrodes, with limits of detection between 10^{-5} and $< 10^{-19}$ mol/l, and the minimum sample volumes needed for analyses (between 0.05 and 1 ml) together with non-destructive measurements also open new vistas in the field of trace analysis. Those ions not detectable with these sensors can be subsequently investigated in the same solution with other techniques, such as atomic absorption.

These new analytical probes are without rival in applications where continuous flow measurements are required, as in the cases of process monitoring, of kinetic measurements, or postoperative monitoring of pH, pNa and pK of blood. Their sturdy, largely maintenance-free construction and broad dynamic response range (3 to $< 10^{-10}$ mol/l), together with the short response time (a few seconds) is especially valuable for industrial applications (process control, waste water monitoring), because subsequently coupled controlling devices can intervene with minimal delay.

Although it is worthwhile to emphasize the significance of these new electrochemical methods by way of introduction, it is not intended to over-rate them here. By no means is the selectivity high enough with all electrodes that it is possible to make a direct, interference-free measurement of the desired ion in the presence of all manner of other ions. Often the composition of the test solution must be known to some degree in order to assess suitable preventive measures (complexation, precipitation or ion-exchange), so that interferences from other ions also sensed by the electrode can be avoided. In practice this seldom leads to great difficulties, since the appropriate step can easily be combined with sample preparation or dilution, so that almost no additional expenditure of time is required.

The possibility of developing special electrodes for higher-valent ions is limited by an unavoidable uncertainty in the EMF measurement. For a ten-fold change in activity the change in potential is ~59mV for monovalent, 29mV for divalent, 19mV for trivalent and 14.5mV for tetravalent ions. The overall measurement accuracy is limited for purely electrochemical reasons to at best 0.1mV (so instrumentation with a substantially more precise readout is of no use). This error is of course of greater importance for higher-valent ions, so that more accurate determinations become diffi-

[1] EMF *E*lectro *M*otive *F*orce

cult. This is not to say that all the higher-valent ions represented in the above survey such as Al^{3+}, La^{3+}, Zr^{4+}, Th^{4+}, etc. can only be determined inaccurately. Since they are all determined indirectly, most via a titration with a reagent which is directly indicated, the usual accuracy of volumetric analyses (0.1 to 0.5%) can be expected.

For the rest there is no general, best method of determination. For each ion, and depending on the particular matrix, there is only one optimum (i.e. simplest, and thus most economical, yielding sufficient accuracy) analytical procedure. Indeed, finding it is precisely the task of the analyst.

1 Fundamentals of Potentiometry

1.1 Electrode Processes

Electrodes are devices with which one can detect the movement and separation of charges occurring at phase boundaries, as well as induce and vary such processes by means of a forced current flow. An electrode can be a piece of some sufficiently inert conductor (such as Pt, Ag, Cu, etc.). If such an electrode is immersed in an electrolyte solution which contains ions of the electrode material, a potential difference will develop between electrode and solution which depends on the activity of this particular metal ion in solution. This is referred to as an *electrode of the first kind.*

An inert electrode, such as platinum (in a solution *not* containing platinum ions) which indicates the oxidizing or reducing power of a specific redox system, is called a *redox electrode.*

If a metal electrode is coated with a thin layer of one of its slightly soluble anion salts (for example Ag/AgCl), then the potential difference depends on the activity of this anion in solution. This is an *electrode of the second kind.* If the slightly soluble metal salt layer contains yet another cation, which also forms a sparingly soluble salt with the anion, but has a larger solubility product than the electrode metal-anion salt (for example $Ag/Ag_2S/CuS$), then the potential difference depends on the activity of this cation in solution. Such a device is an *electrode of the third kind.* Between the electron-conducting material and the electrolyte there can be many other materials, such as semiconductors [1, 2, 3], or complete insulators like glass (glass electrode) and organic compounds (ion exchange and ion-solvating liquid membrane electrodes). At these phase boundaries a potential difference develops which depends on the activity of a specific ion which is present both in the electrode and in solution, and which can migrate easily between the two phases. Such electrodes which selectively sense a specific ion in the presence of others are called *ion-selective electrodes.*

In order to understand these electrode properties, we must refer to an atomic picture of electrode processes. Let us consider the state of material far removed from the phase boundary. In this homogeneous phase (electrode material or electrolyte solution) the opposing forces (attraction and repulsion) of the individual charged or polar particles balance each other overall. The sum of all forces acting on a single particle is zero.

In such an equilibrium situation the cations and anions are uniformly distributed throughout the electrolyte. The water molecules with their large dielectric constant ($\epsilon = 78$) have no long-range orientation, aside from the molecular range of the hydration sphere, so that overall electroneutrality prevails.

Near a phase boundary the overall cancellation of opposing forces on the particles is distorted. An ion in solution, but located near the interface "sees and feels" the electrode, which has entirely different properties than the ions and polar molecules in the bulk of the electrolyte solution. Analogous considerations apply to the electrons or atoms inside the electrode material near the surface. Consequently, a particle in direct proximity to the interface is acted upon by a net force vector. The result of this anisotropy in the immediate vicinity of the phase boundary is the rearrangement or reorientation of those particles "within reach" of this directed force. The goal of such a rearrangement is to reach the lowest energy state in this altered force field. Here "within reach" means a span of a few ionic radii. Thus, there can be a small or large dipole orientation of the solvent molecules near the electrode surface. The oriented dipoles may be thought of as a charged capacitor plate at a distance d from the surface. Thus an *electric double layer* is formed which would behave like a capacitor in an electric circuit, leading electrochemists to coin the quantitative term "double layer capacitance". In electrolyte solutions the solvent molecules are present in large excess and each ion is more or less solvated, so upon approaching (for example through thermal motion) an electrode, the solvation sphere hits the electrode surface first. Thus one can assume as a first approximation, that a monomolecular solvent layer is built up in a stepwise fashion on the electrode surface. The depth of this molecular layer constitutes the so-called *inner Helmholtz plane* (d_{H_i}).

Because of anisotropic interactions in the vicinity of the phase boundary, charged particles, i.e. ions, can be enriched or depleted behind this initial solvent layer relative to the average concentration in bulk solution. The plane in which the majority of these ions can be thought to lie is called the *outer Helmholtz plane.* This excess or deficiency of charge at a distance d_{H_o} induces, in turn, an opposite charge of the same magnitude on the electrode surface, as shown in Fig. 3.

One refinement of this model is especially important for making measurements. One should not imagine that the ions which are enriched in the Helmholtz plane are permanently and rigidly positioned in this plane. Because of thermal motion and collisions with other ions and solvent molecules, at least a portion of this excess charge extends out into the electrolyte solution to some depth. This "smearing out" of the excess charge increases with decreasing ion concentration. What significance does this model have for making measurements? In practice, one often works with a fixed electrode relative to a moving electrolyte solution (for example in flow measurements, rapidly stirred thermostated solutions, etc.). We know from the hydrodynamics of a liquid (electrolyte) moving relative to a fixed wall that, no matter how great this relative motion, a more or less thick layer of liquid is retained due to forces of interaction with the wall. As a rule, the thickness of this stationary layer extends far beyond the outer Helmholtz plane. This is not the case when one goes to very dilute solutions, since here a "tail" of sufficient excess charges exists from diffusion into solution that some of these excess charges get dragged along with the flow. This transport of excess electric charges by definition corresponds to an electric current i_{hyd}, as indicated in Fig. 4. If an electric current flows through a medium with a given resistance R (here the reciprocal of the conductivity of the electrolyte), then according to Ohm's Law, a voltage drop $R \cdot i_{hyd}$ results. Thus, a potential exists due to the movement of a very dilute electrolyte relative to a fixed electrode. This potential is often denoted as the

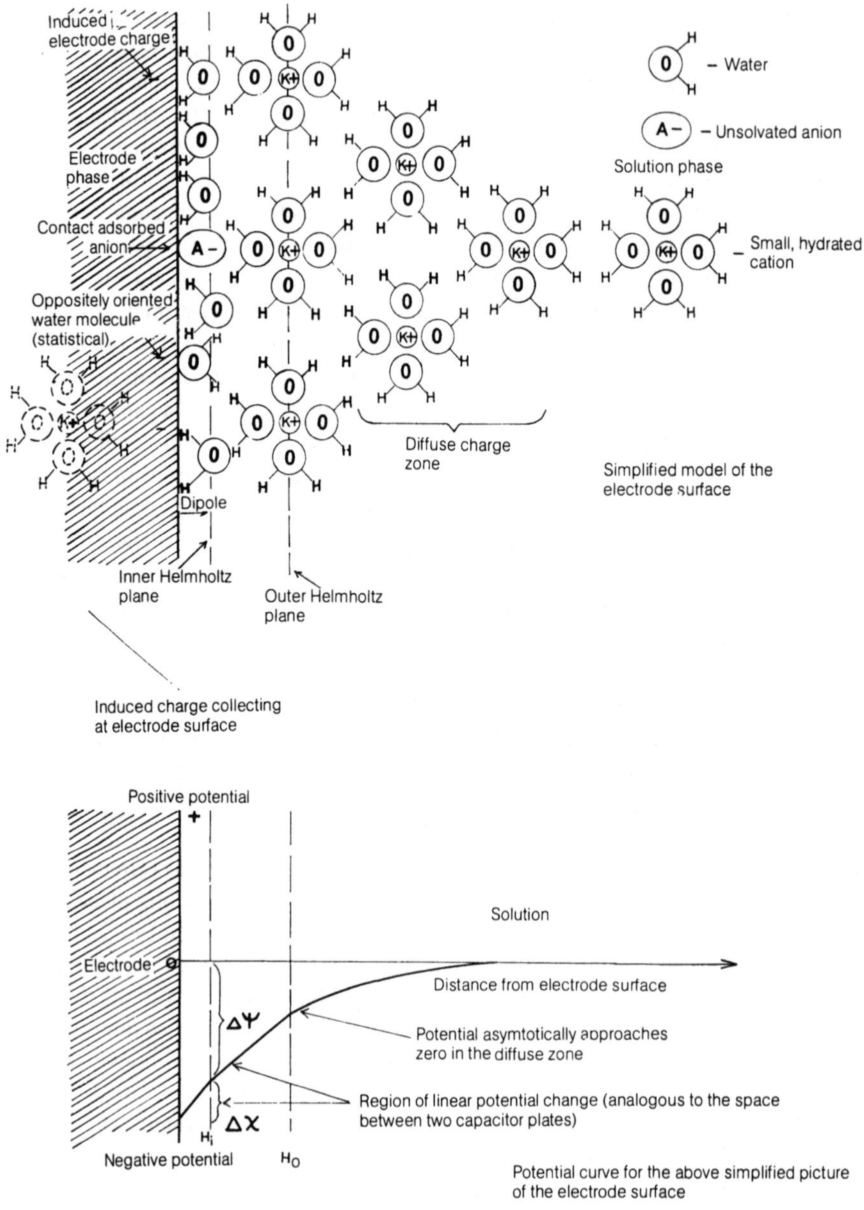

Fig. 3. Simplified model of the electrode surface. (For the sake of clarity all but one of the anions have been left out; the hydrated cations are used to symbolize the excess charges)

electrokinetic or ζ *(Zeta) potential.* It can be annoying for analytical measurements, in that one may find a strong dependence of the data on stirring rate or cell geometry. One can minimize this effect by reducing the diffuse excess charge portion of the Helmholtz layer and at the same time improving the conductivity of the solution,

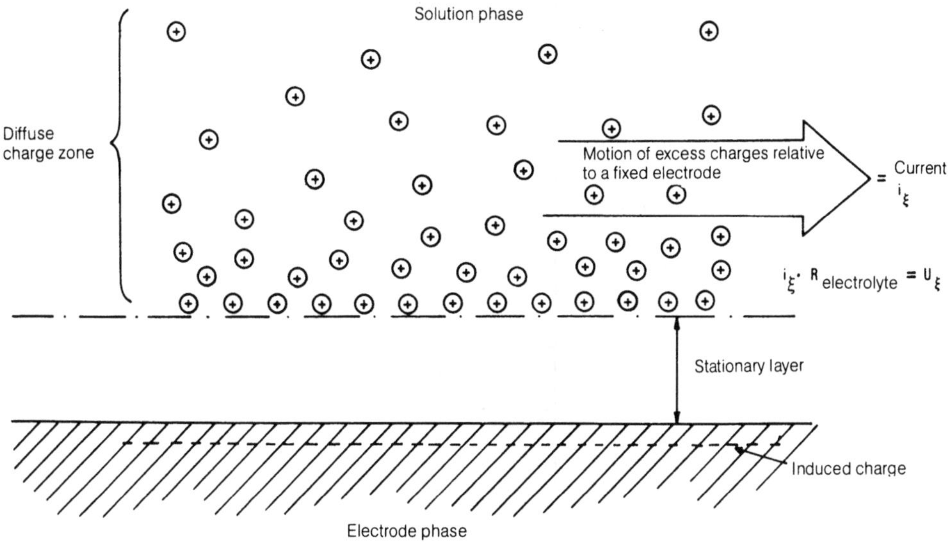

Fig. 4. Schematic explanation of the streaming potential

both of which can be accomplished by the addition of an inert electrolyte (one which does not disturb the indicator electrode).

So far our considerations have only illustrated how a partial charge separation at the electrode/electrolyte interface comes about. In summary, the anisotropy due to dispersion forces makes a partial charge separation possible. There has as yet been no mention of the passage of an ionic species from the solution phase into the electrode phase or vice versa, which a charge transfer through the electric double layer would require. In the first case one refers to an ideally polarizable electrode. The other extreme in this connection is the completely unpolarizable electrode with a great deal of ionic exchange between the two phases. This can be characterized in an equivalent circuit by a short-circuited capacitor (Fig. 5). This means that a current across the phase boundary does not influence the potential across the capacitor or, likewise, the electrode potential. This property is especially important for the measuring technique, since a small current can never be completely avoided in any potential measurement. A well-known example of a polarizable electrode is the mercury-drop electrode of polarography; conversely, the common reference electrodes (Ag/AgCl, Calomel, etc.) belong to the essentially unpolarizable type.

How do the additional processes at the electrode/electrolyte phase boundary manifest themselves to the electrochemist? In the first instant that the electrode is immersed in solution, neither the electrode nor the solution is charged at all; the charge separation process takes place only after initiating this deviation from equilibrium. In aqueous solutions there is a build-up of a water-dipole layer with its associated potential $\Delta\chi$. Simultaneously, a voltage $\Delta\psi$ develops due to an enrichment or depletion of ions in the Helmholtz plane. The ions in solution need not, however, come to a halt before the electrode outside the Helmholtz plane. They can, if they gain energy in the

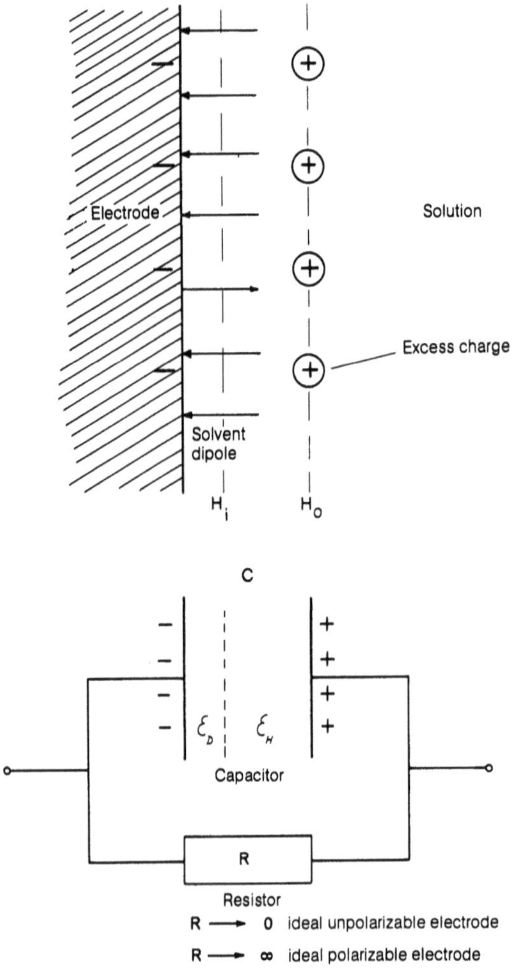

Fig. 5. Electrode surface model and the corresponding equivalent circuit

process, penetrate both the Helmholtz and dipole layers and migrate into the electrode phase, or conversely, an atom or ion in the electrode phase can, through energy gain, enter solution. This "chemical" reaction occurs parallel to the purely electrostatic enriching or depleting process, and is in reality not to be separated from it. In which direction and how far this transfer of charge carriers proceeds is a question of equilibrium. One knows that equilibrium has been reached if, during the course of reaction, energy is released in neither direction; if, therefore, the state of lowest free energy has been reached.

1.2 The Nernst Equation

We know that an electric charge q is associated with an electric field $\mathbf{E}$. The work A which must be done to bring a charge q from infinity to a given point P within an electric field is given by:

$$A(P) = -q \cdot \int_{\infty}^{P} \mathbf{E} \cdot d\mathbf{s} \tag{1}$$

where $\mathbf{E}$ = fieldstrength (= vector)
$d\mathbf{s}$ = path element (= vector)

The integral term is called the electrostatic potential of the point in question:

$$\varphi(P) = -\int_{\infty}^{P} \mathbf{E} \cdot d\mathbf{s}. \tag{2}$$

In the description of an electrode/electrolyte interface, the electrochemist defines a so-called "outer potential" ψ. This is the potential in the vicinity ($\sim 10^{-5}$ cm) of the interface, but still out of range of any image force interactions (in a thought experiment one imagines electrode and electrolyte separated and in a vacuum). The difference between the outer potentials of the electrode and electrolyte solution

$$\psi_{\text{electrode}} - \psi_{\text{solution}} = \Delta\psi_{E-S} \tag{3}$$

is represented by the so-called *Volta potential.*

In order to arrive at the potential difference between the interiors of the two adjacent phases, the layer of partially oriented dipoles which develops at the phase boundary must not be overlooked. The work which must be done for a unit charge to cross a dipole layer can formally be assigned to a surface potential difference $\Delta\chi$. This voltage $\Delta\chi$ must then be added to the Volta potential $\Delta\psi$, which only includes charge separations at various locations in a vacuum, in order to get the total potential difference between the electrode interior and the interior of the adjacent electrolyte solution:

$$\Delta\phi = \Delta\psi + \Delta\chi. \tag{4}$$

This so-called *inner potential difference* is called the *Galvani potential.* In calculating the magnitudes of the various contributions to the work above, all physico-chemical interactions (ion-ion, ion-solvent, etc.), which obviously arise at real interfaces, are excluded. Here only the electrostatic interactions are taken into consideration. When one talks about an *electrode potential,* this expression is to be interpreted as a contraction of the term *electrode potential difference.* It is important to realize that in practice one can measure Galvani potentials to some degree of precision only in the same medium (for example between two copper instrument leads), because it is impossible to distinguish between electrical and chemical energies. It makes no sense, therefore, to measure more precisely than the variations in $\Delta\chi$ in the system (for example at soldered joints, plug connections, etc.).

Calculation of the electrode potential at thermodynamic equilibrium is simple: one can define a *chemical potential* μ, analogous to the *electrostatic potential* ϕ, as the work which must be done or is set free if a mole of a given type of particle is to be brought from infinity into the interior of some uncharged phase free of any dipole layer coating. All chemical interactions can be determined with this property.

The total work, and thus the electrochemical potential $\overline{\mu}$, is the sum of electrostatic ($z \cdot F \cdot \phi$) and chemical work (μ):

$$\overline{\mu} = \mu + z \cdot F \cdot \phi \tag{5}$$

where $\mu = \left(\frac{\partial G}{\partial n}\right)_{P,T}$ = chemical potential (derivate of free energy wrt # moles)

z = number of electronic charges exchanged per molecule

F = Faraday constant (96496 A · s/mole).

The factor $z \cdot F$ takes into account the fact that the definition of chemical potential is based on the mole, while the Galvani potential is defined in terms of the unit charge $q = 1 \mathrm{A} \cdot s$. The Faraday constant arises from multiplying the electronic charge e_0 by Avagadro's number: $F = e_0 \cdot N_A$. The factor z, which indicates the number of unit charges exchanged per molecule, must be taken into account for multivalent ions. Equilibrium between different phases prevails, then, when the electrochemical potential is equal in all phases:

$$\overline{\mu}_{\text{phase 1}} = \overline{\mu}_{\text{phase 2}} = \overline{\mu}_{\text{phase 3}} \text{ etc.} \tag{6}$$

In order to reach this state, it is necessary that material transport across the phase boundaries can occur. This implies that unpolarizable electrodes are in a state of thermodynamic equilibrium (in polarizable electrodes material and charge transport across the double layer are forbidden).

For an electrode reaction between a metal ion M^{z+} with charge z and the electrode metal M.

$$M^{z+} + ze^- \rightleftharpoons M \tag{R1}$$

this equilibrium criterion is thus:

$$\overline{\mu}_{M^{z+}} + z\overline{\mu}_e = \overline{\mu}_M \tag{7}$$

where: $\overline{\mu}_M$ = electrochemical potential of metal M in the metallic phase

$\overline{\mu}_{M^{z+}}$ = electrochemical potential of metal ion M^{z+} in the solution phase

$\overline{\mu}_e$ = electrochemical potential of electrons in the metallic phase.

Considering the aforementioned definition of electrochemical potential and solving the equations for $\Delta \phi_{eq}$ yields the equilibrium Galvani potential ($= \phi_{\text{metal}} - \phi_{\text{solution}}$):

$$\Delta\phi_{eq} = \frac{\mu_{M^{z+}} + \mu_e - \mu_M}{z \cdot F}. \tag{8}$$

An exact calculation of the individual chemical potentials μ_i is impossible because of the complex interplay of the various interaction forces. In thermodynamics one refers instead to a standard state μ_i^0, and describes changes relative to it according to:

$$\mu_i = \mu_i^0 + R \cdot T \cdot \ln a_i \tag{9}$$

where: R = gas constant (8.31 W·s/K · mole)
T = temperature in Kelvin
a_i = activity of ion i (see Appendix).

If one further stipulates that the activities of the metal atoms and electrons are equal to unity in the pure metal phase, then:

$$\Delta\phi_{eq} = \frac{\mu_{M^{z+}}^0 + \mu_e^0 - \mu_M^0}{z \cdot F} + \frac{R \cdot T}{z \cdot F} \ln a_{M^{z+}}. \tag{10}$$

Thermodynamics furnishes us with the relationship between the equilibrium Galvani potential at an electrode and the *activity* in solution of those ions which are able to penetrate the phase boundary, and hence the electric double layer. The difference in standard chemical potentials

$$\mu_{M^{z+}}^0 + \mu_{e^-}^0 - \mu_M^0 = -\Delta G^0 \tag{11}$$

represents the "driving force", or Gibbs free energy ΔG (P = const.), of reaction. If one defines a *standard potential* with $a_{M^{z+}} = 1$, then the standard equilibrium Galvani potential is given by:

$$\Delta\phi_{eq}^0 = \frac{\mu_{M^{z+}}^0 + \mu_e^0 - \mu_M^0}{z \cdot F} = -\frac{\Delta G^0}{z \cdot F}. \tag{12}$$

Thus Eq. (10) becomes:

$$\Delta\phi_{eq} = \Delta\phi_{eq}^0 + \frac{R \cdot T}{z \cdot F} \ln a_{M^{z+}}. \tag{13}$$

This equation represents one form of the classical *Nernst equation.* It is easy to show that for the general case of a redox reaction it looks as follows:

$$\Delta\phi_{eq} = \Delta\phi_{eq}^0 + \frac{R \cdot T}{z \cdot F} \ln \frac{a_{Ox}}{a_{Red}} \tag{14}$$

where a_{Ox} and a_{Red} stand for the activities of the oxidized and reduced forms, respectively.

Without a doubt the most interesting feature of this equation for the analyst is the possibility of being able to explore quickly and directly the activity or concentration of a particular ion i through measurement of the equilibrium Galvani potential change with a suitable electrode. This is the field of *direct potentiometry.*

If the activity of the indicated ion is changed via a titration using a complexing or precipitating agent, then the equilibrium Galvani potential follows this reaction according to the Nernst equation with a logarithmic dependence. If one then plots the equilibrium Galvani potential against the added volume of titrating agent, a more or less sharp jump in the titration curve appears at the equivalence point. This is the field of *potentiometric endpoint indications.*

There are further possibilities for the thermodynamicist. At known concentrations, the *activity coefficients* f can be evaluated from $a = f \cdot c$ (see Appendix). The standard equilibrium Galvani potential difference can be determined by choosing $a_{M^{z+}}$ or $\frac{a_{Ox}}{a_{Red}} = 1$. From this one can ascertain the change in *standard Gibbs free energy* $\Delta G^0 = -zF\Delta\phi^0_{eq}$ and thus using $\Delta G^0 = -RT \ln K$ investigate the *equilibrium constant* of the corresponding electrode reaction.

The derivation of the Nernst equation is based on the assumption of chemical equilibrium. *It is, therefore, only to be applied under equilibrium conditions.* Frequently, too little attention is paid to this restriction. If one theoretically considers the definition of chemical potential:

$$\mu_i = \mu_i^0 + R \cdot T \cdot \ln a_i \tag{15}$$

one can see underlying it the formula for calculating the isothermal, *reversible* work of expansion of an ideal gas:

$$A = \int_{P_I}^{P_F} V dP = R \cdot T \int_{P_I}^{P_F} \frac{dP}{P} \tag{16}$$

$$A = R \cdot T \cdot \ln \frac{P_F}{P_I} + C \text{ (integration constant)} \tag{17}$$

where: P_I = initial pressure
P_F = final pressure.

Thus the Nernst equation is applicable only to *reversible* electrode reactions under equilibrium conditions.

1.3 Potential-Determining Ions

In the derivation of the Nernst equation it is assumed that equilibrium has been established between a solution of some particular ion and an adjacent electrode phase. In order for equilibrium to be reached at all, two conditions must be met. First, reasonably unhindered material transport from one phase to the other and back is necessary. Second, the species in question must be present in both phases in sufficient quantity that the material exchange necessary to reach equilibrium does not change the properties of the corresponding phase so much that its identity is lost.

An example of the first requirement is the so-called *hydrogen electrode.* This is a platinum wire covered with catalytically active platinum black which is exposed to a flow of pure hydrogen gas, so that a continuous layer of absorbed H_2 gas molecules is formed on its surface. When dipped in an electrolyte, this set-up works like an electrode made of hydrogen, at which the following reaction can take place:

$$\underset{\text{(Red)}}{\tfrac{1}{2} H_2} \rightleftharpoons \underset{\text{(Ox)}}{H^+} + e^- . \qquad \text{(R2)}$$

The corresponding Nernst equation reads:

$$\Delta \phi_{eq} = \Delta \phi_{eq}^0 + \frac{R \cdot T}{F} \ln \frac{a_{H^+}}{\sqrt{a_{H_2}}} . \qquad (18)$$

If we include the constant activity of hydrogen gas at constant pressure in the constant $\Delta \phi_{eq}^0$, and consider the definition of pH value (pH $\equiv -\log a_{H^+}$), then we have

$$\Delta \phi_{eq} = \Delta \phi_{eq}' - 2.3 \frac{R \cdot T}{F} \text{pH}. \qquad (19)$$

In fact, this strong pH-dependence of the equilibrium Galvani potential difference can be observed with the above construction. Such a large pH-dependence can even be found with a suspension of Raney nickel in contact with a normal platinum electrode. Attempts with other metals, which are known not to catalyze reaction (R2), yielded irreproducible results. This example serves to illustrate the requirement of an unobstructed phase transfer, which in this case was coupled with a chemical reaction.

A vivid example of the second requirement concerning the identities of the two phases is a bare copper wire immersed in a solution containing Cu^{2+} and Ag^+. In this case the Nernst equation describes only the starting potential according to the reaction

$$\underset{\text{(Ox)}}{Cu^{2+}} + 2e \rightleftharpoons \underset{\text{(red)}}{Cu} \qquad \text{(R3)}$$

with

$$\Delta \phi_{eq} = \Delta \phi_{eq}^0 + \frac{R \cdot T}{2F} \ln \frac{a_{Cu^{2+}}}{a_{Cu}} ; \qquad (20)$$

but from the moment of immersion in solution, the electropositive silver ions preferentially deposit on the copper wire, altering the copper electrode surface. In the end there exists a pure silver surface, which itself can once again be described by the Nernst equation using parameters appropriate for this metal. In the interim period, the value of $\Delta \phi^0_{eq}$ changes from that characteristic of a copper electrode to that characteristic of a silver electrode. Measurements in this transition period do not have a constant $\Delta \phi^0_{eq}$ term, and so are not suitable for analytical applications.

With ion-selective electrodes one avoids this complication as much as possible by conditioning an electrode in the appropriate solution for hours or days before the actual measurement. The appearance of some potential drift when working with extreme concentrations and in solvents other than those used for conditioning can often be traced back to the above process of identity change and corresponding change of the $\Delta \phi^0_{eq}$ term. In this vivid example, at the beginning of the experiment the Cu^{2+} ions participated most in the phase transfer, thus determining the potential, and at the end the Ag^+ ions were potential-determining.

Let us now try to make the somewhat lax description of phase boundary crossing more precise. For this the concept of the experimentally accessible *exchange current density* i_0 must be explained. The equilibrium state of a particular ionic species between two phases (electrode/solution) is reached, if the electrochemical potential $\bar{\mu}$ is equal in both phases. If an uncharged ion-selective electrode is dipped in solution, any ion which establishes an exchange equilibrium with the active electrode phase must cross the phase boundary. The initial direction of charge transfer depends on the chemical potential gradient $\frac{d\mu_i}{dx}$ at the phase boundary. The ions choose whichever of the two phases offers a stable state with the lower energy. If the ions had no charge, then this transfer of species from the phase with the higher chemical potential to that with the lower would continue until the chemical potential had become equal in both phases. This is the case with partition equilibria. Since in our case only one ionic species should preferentially cross the boundary, we are dealing with a charge transfer which increasingly charges one phase in opposition to the other. In the end, the ion migration favored by the chemical potential gradient is so strongly hindered by electrostatic forces, that further migration is stopped and the reverse reaction is promoted because of repulsive forces between particles of the same charge and attraction of the oppositely charged particles left behind. At equilibrium, microscopic ionic streams crossing the interface in both directions balance one another.

The transport of charge carriers across the phase boundary by definition corresponds to a directed electric current $\vec{i}$ or $\overleftarrow{i}$. At equilibrium $\vec{i} = \overleftarrow{i} = i_0$. This current density is called exchange current density. No measurable macroscopic current flows through the phase boundary, but equal and opposite microscopic current densities exist nevertheless. They maintain the charge separation which has built up at the interface to this point, and thus allow the measurement of a constant potential, despite the fact that in every measurement a very small current is drawn. The exchange current density i_0 can be experimentally determined for metal electrodes and apparent values can be estimated for ion-selective electrodes [4, 448]. Its value allows one to infer certain things about the kinetics of an electrode reaction. Let us consider the hydrogen electrode and various electrode materials as an example. The following exchange current densities have been measured:

Material	Reaction	Exchange Current Density i_0
Pt, Pd	H_2/H^+	10^{-3} [A/cm^2]
Hg	H_2/H^+	10^{-12} [A/cm^2]
Pb	H_2/H^+	10^{-11} [A/cm^2]

Considering that in practice an instrumental current of 10^{-10} to 10^{-14} A cannot be avoided in an EMF measurement, it is not surprising that irreproducible results and deviations from the Nernst equation are observed with Hg and Pb. The instrumental current is much too large compared to the exchange current density and, hence, strongly disturbs the equilibrium.

For an ion-selective electrode, the ionic species which *carries the greatest number of electric charges across the phase boundary* is potential-determining. This species is the main contributor to the charge separation at the electrode/electrolyte interface actually measured by the so-called electrode potential.

Consider the following step in our model:

Ion in solution → Ion or atom in electrode phase.

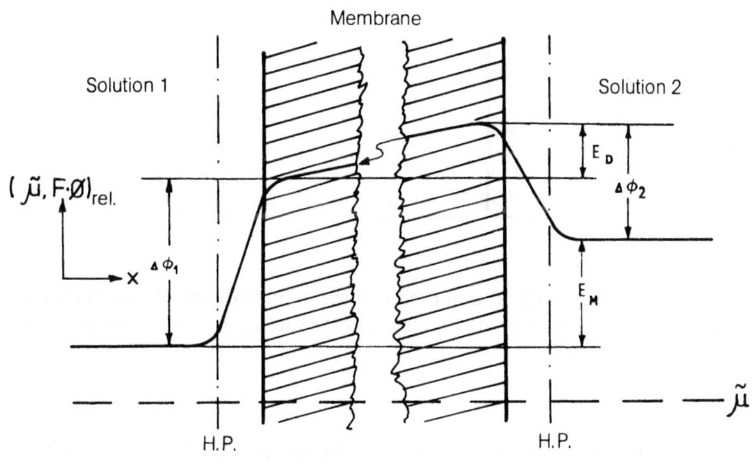

Fig. 6. Schematic representation of equilibrium at an ion-conducting membrane.
$\Delta\phi_{1,2}$ = Galvani potential at interface 1 or 2, respectively
E_D = possible diffusion potential within the membrane (idealized behavior)
E_M = electrode potential with identical reference electrodes

In order for a single ion to cross the phase boundary, it must leave behind its solvation sphere (hydration sphere in aqueous solutions). A specific amount of energy is necessary for this. This energy corresponds to the solvation energy which is set free by dissolution of the ion in question. The solvation energy depends on the charge and radius of the ion. Figure 7 illustrates how the potential energy of a solvated ion depends on

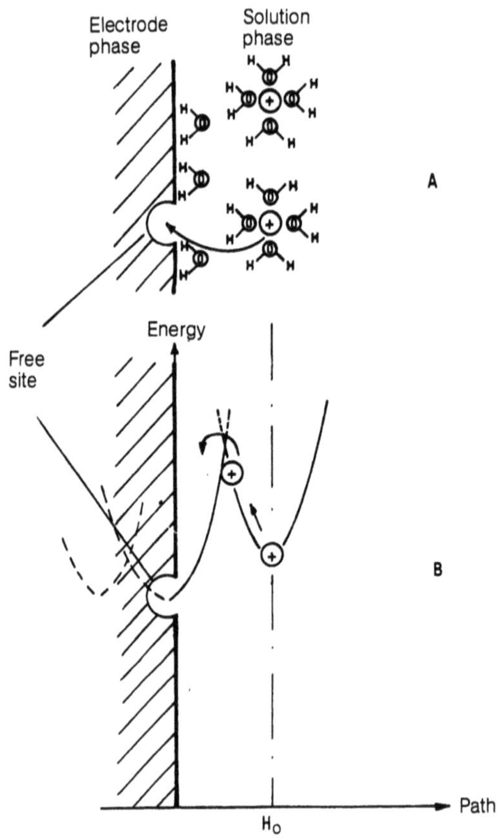

Fig. 7. Schematic diagram of a charge transfer. **A:** Pictorial representation of a cation leaving the Helmholtz plane and occupying a position on the electrode surface. **B:** Corresponding energy diagram

its position relative to the center of this solvation sphere, since it must leave this position in order to cross into the other phase. In Fig. 7 our ion, we suppose a cation, gradually moves out of the stable equilibrium position at the energy minimum. The more it withdraws from the center and nears the water molecules, which it must push aside in order to escape these forces, the more energy it needs. Where does this energy come from? Statistical thermodynamics answers this question by saying: although the majority of ions exhibit a characteristic average energy there are still a few which are much more energetic, just as there are corresponding number of ions which possess less energy. Thus there are a certain number of solvated ions which can move quite far from the center of the hydration sphere. We can assume further that the solvated ions can approach the electrode surface up to a distance d_{H_O} corresponding to the outer Helmholtz plane, as Fig. 7a illustrates. The energy vs. distance diagram below it indicates the "energy pathway" an ion must follow in order to successfully enter the electrode phase.

The number of ions which can cross the boundary per unit time, and thus the exchange rate and exchange current density, are determined by the laws of chemical kinetics. In order to say more about this exchange rate we must consider the energetic situation of the ion in question at the surface of the electrode phase. There it will preferentially be in an energy minimum which lies no higher than that of the solvated ion; otherwise the reverse reaction:

$$\text{Ion in electrode phase} \rightarrow \text{Ion in solution}$$

would occur. Likewise, an ion in the surface layer of the electrode phase is prevented from leaving its stable position, apart from small thermal vibrations, by interaction forces with its neighbors. Here again, we can schematically draw an energy diagram, except here we must differentiate a little more on the right-hand side of the well. The left side rises quite steeply because of the correspondingly large increase in repulsive forces from atoms in the interior of the phase; to be sure, a steep rise can be avoided by overlapping with the potential curves of neighboring lattice vacancies. In this case the ion presently sitting on the surface is able to diffuse further into the phase interior. In estimating the right side of the energy well which extends into solution, one should consider that it can rise more or less steeply, depending on the electrode material and the properties of the Helmholtz layer. If an ion is held by the relatively far-reaching coulomb forces, then the energy function behaves somewhat like that of the solvated ion. A completely different situation exists if the exchangeable ion is held by atomic binding forces, whose effectiveness is limited to a much smaller range. Still other situations exist with ion-dipole interactions or ion-lattice hole interactions.

If we combine the potential wells for both solution and electrode phases, the reaction profile diagram shown in Fig. 8 results. Figure 8 illustrates the situations before and after equilibrium has been reached. During the time in which the electrode/electrolyte phase boundary becomes charged via this ion transfer reaction, the favored ion transfer (directed toward the respective energy minimum) becomes hindered by the increase in free activation energy $\Delta G^{0\ddagger}$ according to: $\Delta G^{0\ddagger} + \beta \cdot F \cdot \Delta\phi$. Until the energy maximum is reached, the transferring ions must do work against the fraction β (also called transfer coefficient) of the total electric field in crossing the inferface, while the reverse reaction is facilitated by the corresponding factor $(1 - \beta)$.

Now we can, in part, answer our question about the parameters which influence the selectivity of an electrode. *The ionic species which "displaces" the greatest number of charge carriers by entering a new phase determines the electrode potential, provided that no other processes are involved, such as diffusion of surface ions into the phase interior. In such a case the slowest partial reaction, as always in kinetics, controls the total reaction.* Both cases are found with ion-selective electrodes. The table below shows some apparent exchange current density estimates for ion-selective electrodes [448].

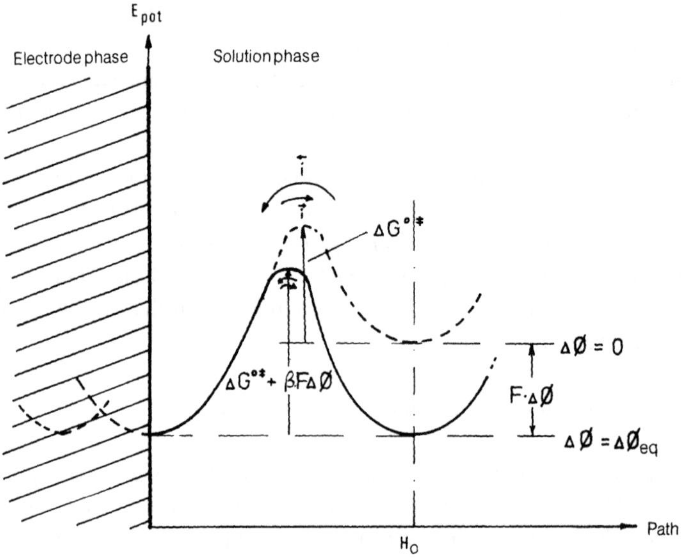

Fig. 8. Schematic reaction profile for a phase transfer – influence of the charge on the activation energy. (At equilibrium the ionic streams $\overleftarrow{i}$ and $\overrightarrow{i}$ are equal, see Fig. 11)

Apparent exchange current density estimates

Electrode	Solution	Exchange current density A/cm²
Ag metal	1 M $AgNO_3$	0.1
Ag_2S	1 M $AgNO_3$	1
Fluoride (LaF_3)	1 M KF; 0.7 M K_2SO_4	$5 \cdot 10^{-5}$
"	1 M KOH; 0.7 M K_2SO_4	$5 \cdot 10^{-6}$
"	0.01 M KF 0.7 M K_2SO_4	10^{-5}
"	0.025 M La $(NO_3)_3$	$7 \cdot 10^{-7}$
$6 \cdot 10^{-7}$ M valinomycin/n-decanol	1 M KCl	$3 \cdot 10^{-7}$
$6 \cdot 10^{-4}$ M valinomycin/n-decanol, not conditioned	1 M KCl	10^{-7}
$6 \cdot 10^{-4}$ M valinomycin/n-decanol	1 M KCl	10^{-5}
$6 \cdot 10^{-3}$ M valinomycin/n-decanol	1 M KCl	$2 \cdot 10^{-4}$
2,7$\cdot 10^{-3}$ M valinomycin/diphenyl ether	1 M KCl	$6 \cdot 10^{-4}$
"	1 M K–picrate	$9 \cdot 10^{-3}$
"	1 M RbCl	$2.7 \cdot 10^{-3}$
"	1 M CsCl	$2 \cdot 10^{-4}$
"	1 M NH_4Cl	$3.4 \cdot 10^{-6}$
"	1 M NaCl	10^{-6}
"	1 M LiCl	$8 \cdot 10^{-7}$
"	0.1 M $MgCl_2$	$2 \cdot 10^{-7}$

1.4 Ion-Selective Electrode Materials

One of the most thoroughly investigated electrode materials is glass [5, 6, 7]. It is not surprising, then, that it has been somewhat in the foreground of this field. More than 60 years ago Cremer [8] described functional pH behavior at a glass membrane. The most well-known pH-sensitive electrode material is without a doubt the so-called MacInnes glass made by Corning Glass Works with the designation 015, whose composition is roughly 72% SiO_2, 6% CaO and 22% Na_2O. One finds with all glass electrodes that stable, reproducible potential values are obtained only after a definite "soaking time" in water. During the soaking process one can also observe that Na^+ ions diffuse out of the glass into solution. Their positions are occupied by H^+ ions from solution [9–12]. The resulting gel layer extends some 10 to 1000 Å into the glass and is of fundamental importance to the electrochemical behavior of the electrode.

As early as 1923 it could be shown [9, 13] that, especially at high pH values, the glass electrode also responds to other monovalent cations such as silver, sodium and potassium ions. In 1934 Lengyel and Blum [14] systematically investigated the connection between glass composition and electrode function, and ascertained that this behavior could be enhanced with certain admixtures of trivalent metal oxides (chiefly Al_2O_3). Thereafter people have endeavored to keep this "error" small in order to avoid additional corrections to the pH measurements. It was found that this error was lessened by replacing the Na_2O glass component with Li_2O [15]. Accordingly, new types of glasses were smelted for which noticeable errors first appeared at pH 14, and even then only at high sodium ion concentrations.

Through systematic investigations of the glass electrode [16–20] it was learned that, above all, the surface structure of the electrode material directly at the solution boundary influences the selectivity. In the interest of the greatest possible selectivity, only the ion to be sensed should have the ability to be reversibly absorbed and released by the electrode material. Investigations by Eisenman et al. [21] showed that the selectivity of glass electrodes is determined by the equilibrium constant of the reaction:

$$\text{Ion}_{\text{solution}} + \text{Ion}^*_{\text{electrode}} \rightleftharpoons \text{Ion}^*_{\text{solution}} + \text{Ion}_{\text{electrode}} \qquad \text{(R4)}$$

where Ion = ion to be sensed
Ion* = ion of electrode material

as well as by the ionic mobility of the ion concerned in the gel layer. In this regard it is not surprising that the pH glass electrode is one of the most selective electrodes, since the radii of a proton and any other ion differ by many orders of magnitude. How much tunneling effects, which are observed in hydrogen bonding, participate in the proton phase transfer in not known. Also yet to be explored is the extent to which the three distinct types of strongly bound protons formulated by Nikolsky et al. [22] represent different proton vibrational modes within the short-range ordered structure of water or modes inside the potential barrier at the electrode surface.

One can thus explain the proton-selective behavior of the electrode material, or more precisely of the glass gel layer, in that the proton has only a relatively small activation energy to overcome for phase transfer; specifically, it can "tunnel" through this energy barrier.

The necessity of "soaking" new glass membrane electrodes (conditioning) arises from the requirement that the electrode phase contain the ion to be measured (here H^+) in the most buffered form possible (in order to avoid a change in identity, μ = const.). Silanol groups can be formed on the surface through the exchange of alkali ions for protons according to:

$$\geqslant Si{-}OH_{surface} \rightleftharpoons \geqslant Si{-}O^-_{surface} + H^+_{solution}. \qquad (R5)$$

The extent of this dissociation determines the amount of charging of the electrode relative to the solution and thus the electrode potential [23]. If the type of glass is too hygroscopic, the gel layer thus formed will be too similar to the solution phase. According to our more kinetically-oriented model, such a glass then loses the activation energy barrier responsible for differentiating ions. At the same time it loses the ability to make the phase transfer of charge carriers possible on the grounds of a difference in the two energy minima. As a result the pH or ion-sensitive behavior will be lost if the electrode is too hygroscopic. Nernstian pH-behavior can only be observed with glasses which absorb an intermediate amount of water [24]. The properties of non-silicatious glasses such as $Na_2O{-}MgO{-}P_2O_5$ or $Na_2O{-}GeO_2$ and $BaO{-}B_2O_3$ have also been investigated [25–27]. Truesdell and Pommer [28] smelted a phosphate glass with 66.7% P_2O_5, 16% Fe_2O_3, 6.1% Na_2O, 5.6% Al_2O_3 and found a sensitivity to a few alkali and alkaline earth ions in the order:

$$Ba^{2+} > Sr^{2+} > Ca^{2+} > K^+ > Na^+ > Mg^{2+}.$$

A potassium-barium aluminum silicate glass showed an ion-sensitive behavior for the following series:

$$Ba^{2+} > Sr^{2+} > Ca^{2+} > Mg^{2+} \quad [29].$$

A susceptibility for divalent ions could be shown with electrodes made with the help of natural "glasses" (for example, obsidan, tuff, tektite) [27]. Unfortunately they are not all specific, giving a mixed potential determined by all mono- and divalent ions present in the test solution.

In this connection the concept of a mixed potential requires a more precise explanation. In the simplest case, two competing electrode reactions give rise to a mixed potential which lies between the independently measured equilibrium potentials of the individual reactions. This deviation from equilibrium potential has the same effect on the individual reactions as an additional, externally applied overvoltage, thus inducing a current and, hence, a material flow in a specific direction. The relationship between applied overvoltage and corresponding current flow is described by the current-voltage curves readily experimentally available in most cases. The establishment of a mixed potential is best graphically represented by a superposition of the current-voltage curves of the contributing electrode reactions. Figure 9 shows such a graph for the case of a silver electrode at which oxygen can be reduced as a second reaction (next

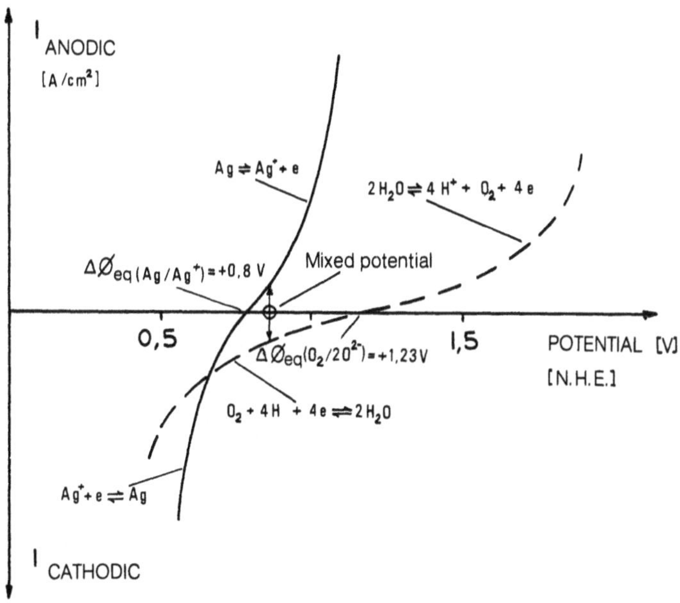

Fig. 9. Establishment of a mixed potential at a metal electrode (the influence of oxygen is exaggerated here). At the measured mixed potential the anodic silver oxidation current is exactly equal to that of the oxygen reduction

to $Ag^+ + e^- \rightleftharpoons Ag$), since its equilibrium potential is more positive than that of the silver electrode. Indeed, in this example its influence on the potential of the silver electrode is slight because its current-voltage curve is quite flat due to inhibiting kinetic factors (corresponding to a small exchange current density). It is easy to see from this illustration that the mixed potential always lies closer to the equilibrium potential of that electrode reaction which has the larger exchange current density. For parallel redox reactions at metal electrodes (corrosion processes for example) Wagner and Traud [30], Kimball and Glassner [31] and Bockris [32] have established, with rather drastic simplifications, a relationship between the mixed potential and the electrode kinetics as well as the equilibrium Galvani potentials, confirming the conclusion reached from analysis of Fig. 9.

In dealing with the establishment of a mixed potential at an ion-exchanger, as in many ion-selective electrodes, it is also important to consider the influence of the exchanging ions. In this case the constantly flowing, equally large streams of different types of ions crossing over the phase boundary can continually change the composition of the surface of the ion-exchanger until a stationary state has been established [33]. At pH-sensitive glass membranes Baucke [23] was able to plot the lithium concentration profiles of a lithium-aluminum silicate glass with a resolution capacity of less than 50Å using an ion-sputtering technique. Following contact with an aqueous electrolyte the lithium concentration (as expected) decreased due to exchange of the protons from the interior of the glass phase to the exterior. The exact concentration profile depends on the electrolyte and it changes reversibly, although on time scales which are long relative to the time needed to establish a stable potential. More recent measurements on glass membrane electrodes show that the glass membrane does not exhibit a uniform resistance; in crossing the gel layer of the dry glass there must be regions with 10^3 times greater resistances than in the glass interior [34, 35]. Baucke [23] was able to localize these regions at about 100Å within the surface. In such regions the lithium concentration differs from that inside the dry phase by about 30% to 50%. In contrast to this, the resistance of the outermost gel layer is very small. The large resistance in the transition region originates from the low migration velocity of the protons in the glass lattice. These new findings must influence the previous way of looking at membrane potentials [36, 37] as a mixture of Donnan and diffusion potentials, since according to these measurements there is no uniform state of diffusion through the entire membrane. The goal of searches for new types of glasses is to find electrode materials selective for each monovalent ion (except the hydrogen ion). These ions do not differ much in radii or, considering their charge, in their resulting field strengths. Thus the selective transfer of a specific ion from solution into the glass gel layer and the associated selective exchange process can only be achieved to within the limits of these small differences. In addition, the size of a potential "location" for a transferable ion from solution varies within certain limits inside the glass structure. All these factors weigh against a greater selectivity. In spite of this, a very good specificity for Na^+ ions is shown with some types of glass. This is due in part to the high ionic mobility of Na^+ ions in the surface layer as compared to the mobilities of competing ions. From tracer studies [21] it is known that the glass phase actually favors the larger potassium, rubidium and cesium ions. But, in the sodium-electrode for example, these ions remain adhered to the surface because of their low mobilities. Hence, they do not contribute as much to the charging of the electrode as the sodium ions, which can more easily diffuse into the glass interior and thus quickly vacate the surface locations. (Here the effect of the kinetics is obvious.)

Many ion-selective electrodes temporarily respond to interfering ions. If one monitors the potential of a sodium-selective glass electrode in a solution with constant sodium ion activity while there is a sudden change in the potassium ion activity a momentary change in potential of from about 0.5 to 10mV appears for a period of about 0.2 to 10 seconds [21, 38]. This effect can also arise with other electrodes, for example with a general monovalent cation-indicating glass electrode if exposed to a sudden change in the activity of some divalent ion ordinarily not indicated, such as Ca^{2+} or Sr^{2+} [39]. For this reason, caution is required when employing ion-selective electrodes in flow measurements. Usually a temporary potential fluctuation inexplicable at first glance can eventually be traced to this effect (see Fig. 10).

The transient signal due to ions which do not normally interfere can also be explained with the aid of our kinetic model of electrode reactions (Fig. 8). In this model the interfering ions are not indicated because their exchange current density is negligible relative to that of the ions detected. Nevertheless, a new equilibrium state must be reached even by these ions after a drastic change in their chemical potential in the test solution. A specific time, determined by the kinetics of phase transfer, is required for this. The pre-equilibrium current (flowing in the direction of decreasing chemical potential) necessary for the restoration of equilibrium can often initially be many times larger than that flowing later under equilibrium conditions (see Fig. 11). In the capac-

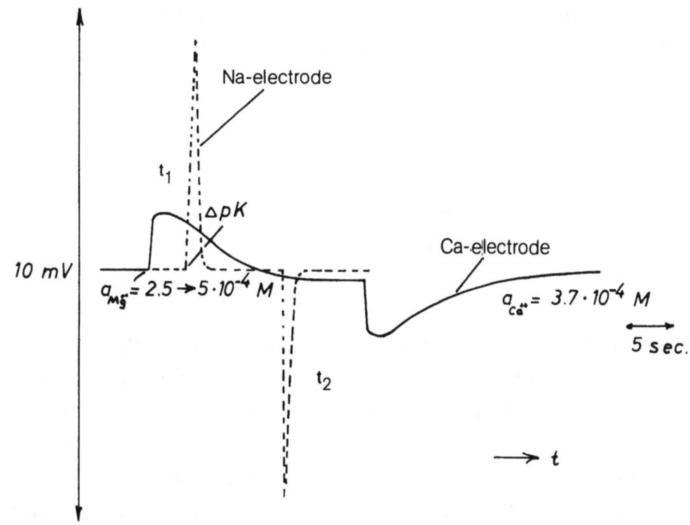

Fig. 10. Transient response due to a sudden change in the interfering ion level of a solution of some indicated ion (t_1 → change in interfering ion activity; t_2 → reestablishment of initial activity)

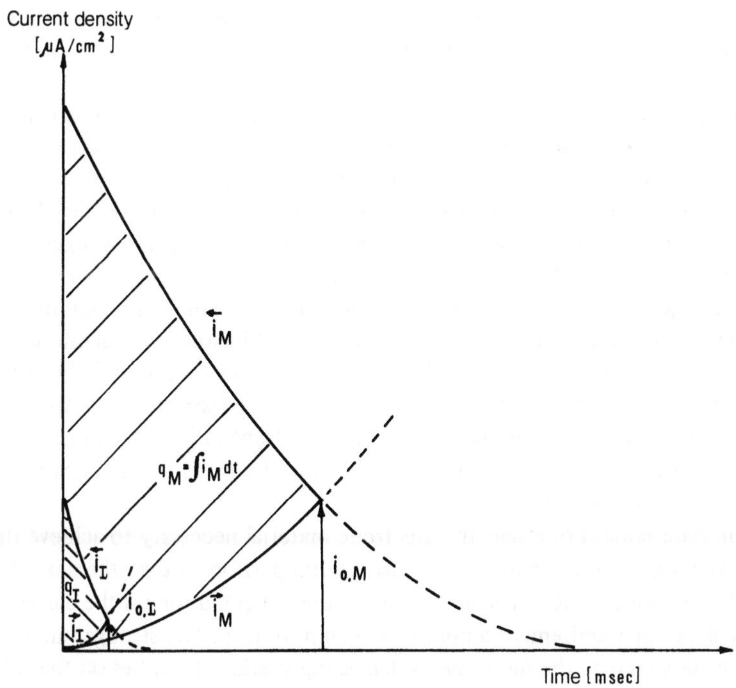

Fig. 11. Schematic representation of the directed currents $\vec{i}$ and $\overleftarrow{i}$ across the interface; the selectivity is given by the ratio of the amount of charge due to the measured ion q_M to that due to the interfering ion q_I.

itor model of an electrode this directed current impulse is associated with a corresponding potential impulse in accord with the laws of electrostatics:

$$U = \frac{\int i \cdot dt}{C} \tag{21}$$

where: U = potential (in V)
i = current (in A)
C = capacitance (in F).

According to this theory this impulse must reverse if the old equilibrium state is again instituted and the ions diffuse back, which is in agreement with experiment. One can explain the observed variation in duration of this temporary response by realizing that various interfering ions at different electrodes have different activation energies. The higher this activation energy is, the fewer the ions which possess the energy required to overcome this energy barrier, the fewer the ions which can cross the phase boundary per unit time, and the longer the restoration of equilibrium lasts. This effect can also be observed with electrodes based on organic exchange materials [40], to be discussed below. With these, however, it continues somewhat longer (about 30 seconds) than with glass electrodes because of the larger activation energies involved. This is also understandable because the selectivity of organic exchange materials rests exclusively on a large inhibition of interfering ion transfer, since the mobilities of ions in such materials do not differ very much in the absence of a matrix of fixed, opposite charges.

The theory of ion-selective electrode operation outlined in Chap. 1.3 can clear up still other as yet unexplained facts. The reduced speed with which the measured ion is indicated in the presence of interfering ions can be explained by a corresponding decrease in the exchange current density of the measured ion, as long as this decrease is not severe enough to noticeably impair the selectivity. The smaller the exchange current density, the longer the time an electrode requires to restore a new equilibrium. The exchange current density can be diminished, for example, at high interfering ion concentrations. The more interfering ions present in solution, the more competition there is for a location in the Helmholtz plane. Such a location must be available for an eventual transfer into and out of the electrode phase (see Fig. 7). If the ions to be detected can occupy a location favorable for phase transfer less often purely on statistical grounds, then this means there will be a retardation in the kinetics of equilibrium and potential restoration.

The main requirement of a selective electrode material necessary to achieve the maximum exchange current density of one ion in comparison to the others is met if the exchange equilibrium (R4) lies on the side of the indicated ion on the electrode phase. With glass this requirement amounts to a compromise. The stronger an ion is bound in the glass matrix, the more the exchange equilibrium (R4) lies on the side of this ion in the electrode phase, and the less mobile it is in the gel layer. The more mobile an ion is in the gel layer, the less favorable its equilibrium lies.

With other electrode materials these conditions are not quite so unfavorable. For example, one can look for *crystals with imperfections.* Because of the uniform size of ionic sites in crystals they can absorb only a few types of ions, and these can diffuse further into the crystal phase via the defects. In order to be useful as electrode materials, these crystals must also exhibit a sufficient ionic conductivity at room temperature. Until now, only a few fluorides of the rare earths (LaF_3) and the silver halides have served as suitable electrode materials.

As the behavior of type 2 electrodes would suggest, various slightly soluble polycrystalline precipitates were tested as electrode materials. Through the application of high pressures, mechanically stable solid state electrodes can be formed, or some supporting material such as silicon rubber or latex can be used for stabilization. With the exception of the silver halides, the selectivity of these "precipitate electrodes" in general is not very great. The silver halide behavior is a consequence of the good solid phase silver ion conductivity observed in these halides (comparable to the sodium-selective behavior of the glass gel layer in the corresponding electrodes). One finds a more or less Nernstian dependence of electrode potential on the activity of the ions which constitute the precipitate (for example $Ba^{2+}SO_4^{2-}$) with the predominantly ionically bound precipitate compounds [41]. Chelated precipitates, however, (such as nickel diacetyldioximine, etc.) with non-polar bonds only occasionally show any effect [42]. With our model this can be explained because the surface potential which must be overcome by an ion in the electrode phase before it can enter solution rises too steeply due to the covalent bonds which must be broken. The resulting activation energy is so high that only a very small exchange current density can be achieved. This current density is only sometimes (when the electrometer input current i_m is less than the exchange current density i_0) sufficient for a stable potential (as illustrated by the hydrogen electrode used with noncatalytic electrode materials).

Fluid or solidified *organic ion exchange materials* are also employed as electrode phases, especially the highly specific ion-solvating molecules. Examples of the latter include some antibiotics [43] and macrotetrolides [44], as well as the synthetically produced so-called "crown" compounds. These cyclic molecules (see Fig. 13) have a cavity which can be "tailor-made" specifically for the trapped ion. The stabilization of an ion from the aqueous phase arises from ion-dipole interaction forces with polar oxygen ligands which extend into the cavity of such molecules. The kinetics of charge transfer is not as strongly hindered by the partially ionic forces as by purely covalent bonds. The mobility of each type of ion in the organic phase is indeed larger due to the absence of a fixed lattice of opposite charges (SiO_4^{4-}-framework in glass, La^{3+} resp. adjacent halide-lattice sites in LaF_3 single crystals resp. in Ag-halide precipitates), but is no longer very different for the various types of ions. Hence this type of electrode loses the ability to distinguish ions on the basis of ionic mobility. The selectivity, then, can arise solely through the ion exchange equilibrium at the interface. This is where the differentiation of specific ionic types must occur.

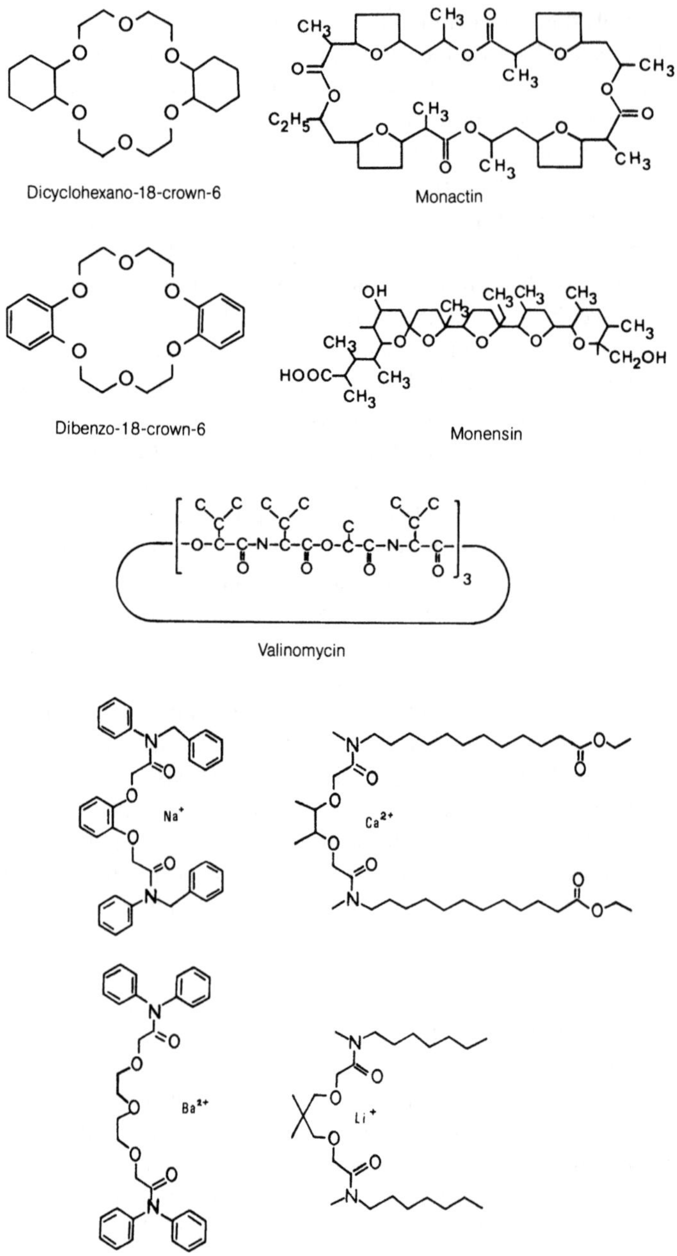

Fig. 12. Structural formulas of some compounds with specific cation-solvating abilities [136]

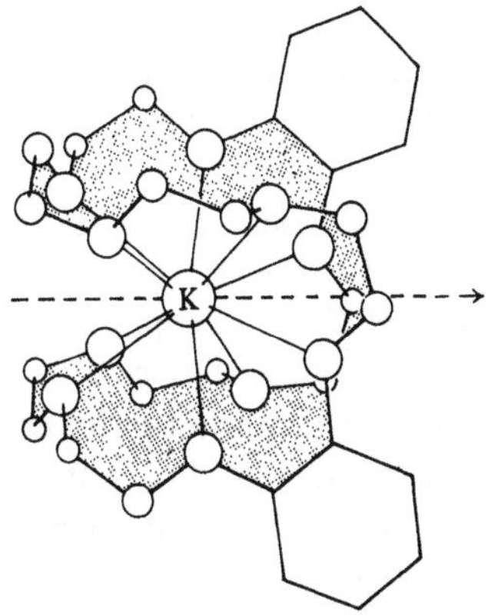

Fig. 13. X-ray structure of the K^+-dibenzo -30- crown- 10 complex [45]

1.5 The Potentiometric Selectivity Coefficient as a Quantitative Indication of the Electrode Selectivity

Since the analyst is usually interested only in the concentration or activity of one particular ionic species, any signal from other ions results in an interference. This type of first-order interference can be approximately described by an empirical Nernst-equation formulated by Nikolsky [46]:

$$\Delta\phi_{eq} = \Delta\phi'_{eq} \pm \frac{RT}{z_M F} \ln \left[a_M + \sum_{a_I} k^{pot}_{M\text{-}I} (a_I)^{z_M/z_I}\right] \qquad (22)$$

in which: (+) is to be used with cations, (−) with anions

a_M = activity of the measured ion
a_I = activity of the individual interfering ions
z_M = valence or change in charge of the measured ion
z_I = valence or change in charge of the interfering ion
$k^{pot}_{M\text{-}I}$ = potentiometric selectivity coefficient (measured ion – interfering ion).

The term selectivity constant has been used for the coefficient $k^{pot}_{M\text{-}I}$, although strictly speaking it is not constant, and with ion-exchange and ion-solvating electrodes $k^{pot}_{M\text{-}I}$ is a function of the absolute ionic strengths of the measured and interfering ions. With the calcium-electrode in dilute solution (< 0.1M) a selectivity coefficient $k^{pot}_{Ca\text{-}Na} \approx 2 \cdot 10^{-4}$ [47] was found for the interference of sodium ions, but in 6M NaCl–$CaCl_2$ a less favorable $k^{pot}_{Ca\text{-}Na} \approx 0.3$ [48] results. In addition, this coefficient also depends on the method used for the determination. In a later section, the methods recommended by an IUPAC commission will be discussed in greater detail.

In contrast to the liquid membrane electrodes, the selectivity coefficients of the exclusively fluoride or silver ion-conducting solid membrane electrodes are clearly determined by the ratio of the corresponding solubility products, and are thus independent of the respective electrode manufacturers.

A small value of the selectivity coefficient defined in Eq. (22) means a high specificity for the measured ion compared to the interfering ion specified by the subscript. If the selectivity coefficient equals 10^{-3} for example, then the measured ion is detected 1,000 times more sensitively than the interfering ion; i.e. a 1,000-fold larger activity of an interfering ion of the same valence relative to the measured ion produces an equilibrium Galvani potential exactly equal to that of the measured ion alone. This corresponds to an error of 100% (twice as large a reading). This also points out how important it is to understand the "selectivity constants" for possible interfering ions specified by the manufacturers before using ion-selective electrodes.

The elimination of interferences starts with the selection of a suitable electrode. Accordingly, one selects from the manufacturer's large supply those ion-selective electrodes which exhibit the smallest coefficient for the most strongly interfering ion. Indeed, caution is required for this because a few manufacturers prefer to state the inverse of this value. If the phrase "more sensitive than" is used, it always refers to a specification of the inverse of the corresponding selectivity coefficient. One must also realize that the numerical values can only be correct reference values when detailed conditions are included. In critical cases, suppose the determination of calcium in blood using electrodes with relatively large $k^{pot}_{Ca\text{-}Na}$ values, the electrodes under consideration should be tested under conditions identical to those of the experiment before ordering. In certain cases the selectivity can be determined with only two measurements following the procedure specified in Chap. 5.9. If the selectivity ratio has a value of about 1, the activities of the measured and interfering ions contribute in equal amounts to the electrode potential. This is not always a drawback; with electrodes for determining water hardness just such behavior is desired.

If a selectivity ratio $k^{pot} \geq 1$ is specified, the ion in question is indicated more favorably than the ion for which the electrode is named. This is the case with almost all Ca-electrodes, for which as a rule Zn^{2+} is indicated more sensitively by a factor of three.

If the test solution contains too high a concentration of an interference ion, one need not give up. An attempt can first be made to selectively lower this high concentration. This can be done by specific precipitation (SO_4^{2-} with Ba^{2+} for the NO_3^- electrode), specific complexation (Al^{3+} with citrate for the F^- electrode), specific ion-exchange (Cl^-, HCO_3^- for the NO_3^- electrode), etc. Frequently the hydrogen ion interferes either by being sensed directly, or indirectly by complexing the measured ion ($F^- + H^+ \rightleftharpoons HF + F^- \rightleftharpoons HF_2^-$ etc.). This interference is easily overcome, however, by shifting the pH with some buffer system as far alkaline as the stability of the other ions in solution allow. Some manufacturers offer "diluting solutions", with which the sample solution is diluted (1:1) before making the measurement. Along with a suitable buffer system these solutions often contain complexing agents to bind all possible interfering ions and to avoid the precipitation of amphoteric hydroxides. Special techniques for the suppression of interferences will be given with the description of the individual electrodes and in the section on applications.

2 Electrode Potential Measurements

2.1 Reference Electrodes

How does one actually measure the potential difference between an electrode and electrolyte, so far only theoretically discussed? A potential-measuring device (voltmeter) has two input terminals. The measuring electrode can be connected with one terminal, but how can one "shunt off" the potential inside the electrolyte solution to the other terminal of the device? There is no other alternative but to immerse a second conductor in the test solution as a "shunt" electrode and connect it to the second input terminal of the voltmeter. However, at this newly formed phase boundary ("shunt" electrode/test solution) the same physical and chemical phenomena arise (orientation, charge separation, charge transfer) as those at the working electrode, whose individual, absolute magnitude one actually wishes to measure. At this second interface we get an additional Galvani potential $\Delta\phi_R$. What we read off of the voltmeter is then the sum of several Galvani potentials. A set-up of this type – two electrodes in a common electrolyte – is called an *electrochemical cell*; similarly, *one* electrode constitutes an *electrochemical half-cell.*

Figure 14 illustrates that the sum of at least *three* potential differences is included in each electrochemical potential measurement, i.e., for each EMF measurement. In addition to the two Galvani potentials at the two electrode/electrolyte phase boundaries, $\Delta\phi_{\text{indicator electrode}}$ and $\Delta\phi_{\text{shunt electrode}}$, yet another arises in any measuring circuit, namely that at the point of contact between two different metals $\Delta\phi_{M/M'}$. In order to measure anything, one is forced to complement the working electrode with a second, different type of electrode – a *reference* electrode. A separation of the EMF measured by the voltmeter into working and reference electrode portions is not possible. One must be satisfied with measurement of the *change* in Galvani potential of a particular electrode and take care that the Galvani potential of the reference electrode as well as that at the point of contact of the two different metals changes as little as possible. Only a few of the large number of possible reference electrodes meet this requirement. *One should beware of valuing only the indicator electrode alone for analytical results. Manufacturers of ion-selective electrodes and experts in the field agree that the main source of trouble in EMF measurements lies precisely with the reference electrode!*

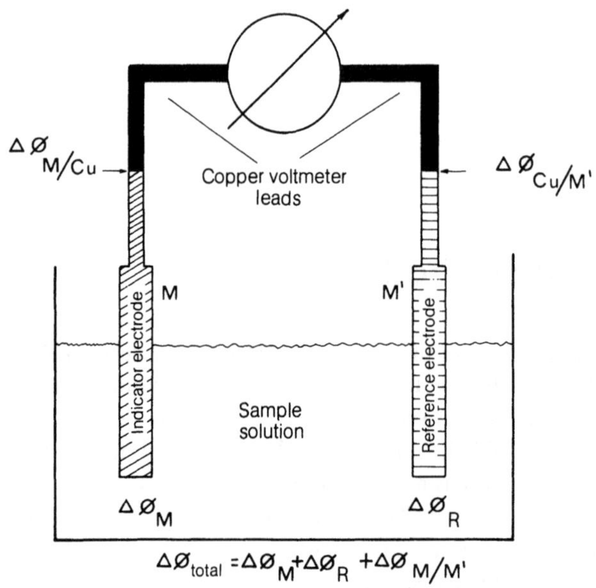

Fig. 14. Schematic representation of the individual potential differences involved in every EMF measurement

2.2 The Standard Hydrogen Electrode as a Primary Reference Electrode

Since a small current in the measuring circuit is unavoidable in any measurement due to the detection technique (potentiometer or direct reading voltmeter), the reference electrode Galvani potential should respond to this current as little as possible. This requirement is met with unpolarizable electrodes with large exchange current densities i_0 (see Chapter 1.3). The so-called *standard hydrogen electrode* has been internationally (*I*nternational *U*nion of *P*ure and *A*pplied *C*hemistry, IUPAC) adopted as the standard reference electrode, since it is sufficiently unpolarizable and can be reproducibly prepared. Its construction has already been described in Chap. 1.3. The equilibrium Galvani potential of the standard hydrogen electrode (at $a_{H^+} \equiv 1$) is, by convention, set equal to zero at *all temperatures.* Further, the standard hydrogen electrode should always be written to the left according to the formal notation of an electrochemical cell. For the case of a silver wire in a silver nitrate solution the following is written:

$$(Pt)H_2\,(p = 1\text{ atm}) \mid H^+_{(a_{H^+} = 1)};\; Ag^+_{(a_{Ag^+} = x)} \mid Ag.$$

In such an electrochemical cell, the experimentally measured potential between the platinum wire of the standard hydrogen electrode and the silver wire represents the magnitude of the so-called *"single electrode potential"*. In reality it is still a sum of potential differences. If the activity of the potential-determining ion is unity, $a = 1$, then the *standard electrode potential* ϵ^0 is measured. The polarity of the right electrode determines the sign, since by convention the potential of the left half-cell is always subtracted from that of the right (for the standard hydrogen electrode $\epsilon^0 = 0$). In practice this means that the reference electrode is connected to the circuit ground (often labeled (–)-terminal). In the above notation of an electrochemical cell, the symbol Ag^+/Ag at the right-hand electrode represents the reaction:

$$Ag^+ + e^- \rightleftharpoons Ag, \qquad (R6)$$

thus a reduction, to which the experimentally measured electrode potential E corresponds. Knowing the single electrode potentials ϵ^0, which can be found in tabular compilations [49,449], the cell voltage of any arbitrary electrode combination can be computed according to the convention: standard electrode potential of the right electrode minus standard electrode potential of the left:

$$\epsilon^0_{\text{right}} - \epsilon^0_{\text{left}} = E. \qquad (23)$$

2.2.1 Preparation

A platinum foil about 1 cm^2 is attached with aid of a spot welder to a platinum wire sealed in pyrex glass to increase the electrode surface area. This foil may be rolled or folded to keep the volume occupied small. Either inside or outside the glass housing the other end of the platinum wire is connected to a regular bare copper wire. After degreasing the platinum foil surface with benzene or another suitable solvent, chemical washing with hot HNO_3 (1:1) and subsequent thorough rinsing with distilled water, the electrode is immersed in a solution of hexachloroplatinic acid [3.5%, with the addition of 0.005% $Pb(CH_3COO)_2$]. It is then electrolyzed with a cathodic current density of 30 mA/cm^2 for five minutes with vigorous stirring (no gas evolution at the cathode). This is done in an H-cell (so that the anodically formed Cl_2 does not migrate to the cathode) with the help of a second platinum electrode as anode [50]. The addition of a small amount (~ 0.01%) of lead acetate to the solution prolongs the life span of the electrode (time until deactivation) [51]. More recent investigations indicate that for a dispersion of any material catalytic for the reaction:

$$\tfrac{1}{2} H_2 \rightleftharpoons H^+ + e^- \qquad (R7)$$

(such as Raney-Nickel) the occasional contact of this material with the platinum foil electrode is sufficient to produce electrodes which agree with one another to within better than 0.01 mV [52].

The necessity of having absolutely pure hydrogen on hand can be overcome by a simple trick (although thanks to the commercially available palladium diffusion purifiers this is no longer a big problem): One can electrolytically generate the hydrogen itself at the platinum electrode with the help of an additional electrode and a current source [53] (Fig. 15). Any accompanying polarization overvoltage is neglected in this case. The total pressure of the evolved gases is equal to the outer air pressure plus the average hydrostatic pressure of the column of solution between the electrode and the solution surface. To determine this pressure reliably one should choose a stable electrode construction which allows direct observation of the inner geometry.

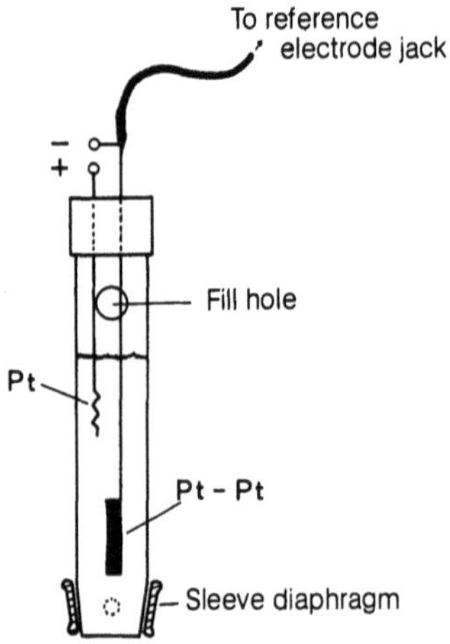

Fig. 15. Principle of the hydrogen-generating reference electrode according to Giner [53]

A further correction (see Appendix) accounts for the change in partial pressure of water vapor with temperature. This pressure must be subtracted from the total pressure in order to get the true partial pressure of pure hydrogen, on which the agreement between different electrodes is based. If a hydrogen electrode and an ion-selective measuring electrode are used in the same electrolyte without an intervening salt bridge, this is called a measuring circuit without liquid junction and the electrode potential is labeled ϵ^0. An electrochemical cell without liquid junction yields the most accurate measurements. Such cells should always be employed unless the composition of the measuring solution affects the equilibrium Galvani potential of the reference electrode.

2.2.2 Characteristics

As shown in Chapter 1.3, the equilibrium Galvani potential of this electrode depends upon the pH of the solution according to:

$$\Delta \phi_{eq} = \Delta \phi'_{eq} - 2.3 \frac{R \cdot T}{F} \text{pH}. \tag{24}$$

For a hydrogen ion activity of $a_{H^+} = 1$ (thus pH = 0) the following equality applies[1]:

$$\Delta \phi_{eq} = \Delta \phi'_{eq} = 0. \tag{25}$$

This holds for all temperatures. The choice of this electrode as a primary reference electrode is due in large part to its excellent reproducibility. Two electrodes constructed using different procedures can still agree in potential to within less than 0.01 mV [54, 55]. For the most precise EMF measurements (such as a control to determine if an electrode can be described exactly by the Nernst equation, measurement of a standard electrode potential and its temperature dependence, etc.) a hydrogen electrode is usually used in a *cell without liquid junction* (without a salt bridge and the associated *liquid junction potential*). It is used less frequently for routine measurements because it is somewhat cumbersome to use. Poisons for platinum catalysts such as CN^-, S^{2-}, SCN^-, etc. interfere when in direct contact with the test solution by decreasing the reversibility of the electrode reaction. Sensitive redox systems, such as nitrophenol, which can be oxidized or reduced with the H_2/H^+ system in the presence of a platinum catalyst also interfere if no precautions are taken to keep these materials far from the electrode surface. This can only be accomplished by means of a salt bridge in the construction of a *cell with liquid junction*. Unfortunately, the resulting liquid junction potential diminishes the precision of the measurement.

2.3 The Liquid Junction Potential

In a cell with liquid junction the salt bridge electrolyte must not mix with the measuring solution too fast or the construction will be useless. This mixing can be slowed down by keeping the area of contact between the two phases small. To this end, one or more capillary connections, ceramic or sintered glass diaphragms, agar-agar plugs, electrolyte-coated sleeve diaphragms, etc. can be employed. The nature of this sample solution/salt bridge contact region has a large influence on the reproducibility of the corresponding EMF measurement because it is here that the so-called liquid junction potential arises. For this reason the details of this process will be described in greater depth. In order to carry out exact measurements with ion-selective electrodes a correspondingly constant liquid junction potential must be assured. *From the standpoint of reproducibility, this contact zone is the weakest link in the chain.*

In the electrochemical shorthand the contact zone of two different electrolyte solutions (also, for example, for the same electrolyte at different concentrations) is represented by a double line ||. This is in contrast to the electrode/solution interface which is symbolized by a single line |. Thus for a cell with liquid junction made up of an ion-selective measuring electrode M in contact with a solution containing the corresponding M^+ ions complemented with a standard hydrogen electrode we would write:

$$\text{(Pt)}\ H_{2\,(p\,=\,1\ \text{atm})}\ |\ H^+_{(a\,=\,1)}\ ||\ \text{salt bridge (3}\ M\ \text{K Cl, for example)}\ ||\ M^+_{(a\,=\,x)}\ |\ M.$$

[1] More recent calculations show that the absolute standard potential ϵ^0_{abs} (H^+/H_2) must be about 4.31V (450, 451).

2.3.1 Origin

Let us consider as a simple example a 1 M KCl solution and a 1 M HCl solution separated by a diaphragm. The two solutions contact one another only inside the channels of the diaphragm. At these contact points large K^+ and H^+ ion concentration gradients, dc/dx, develop because one side contains none or only trace amounts of the other cation. This concentration gradient, or more precisely this activity gradient, is the driving force behind the diffusion of both cations which commences upon contact of the two solutions. The speed and depth of this diffusion depends on the activity gradients (equal in this case) and the mobilities of each of the ions in the respective phases.

It is known that the migration velocity of H^+ ions in water is about 5 times that of K^+ in water. Therefore the H^+ ions will migrate about 5 times faster into the KCl solution (per unit time this means 5 times the number) than the K^+ ions into the HCl solution (Fig. 16). In our example the Cl^- concentration is the same in both solutions, so no net

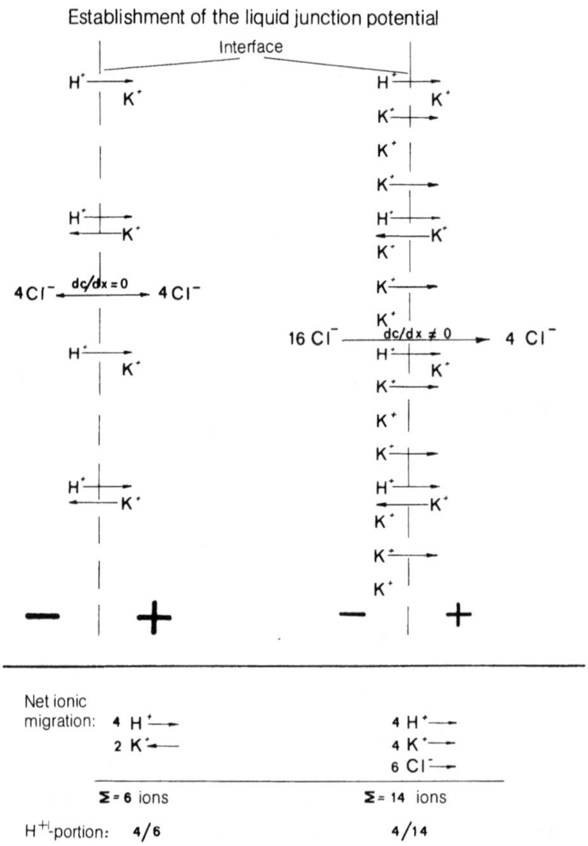

Fig. 16. Schematic representation of the establishment of a liquid junction potential at an HCl || KCl (equal concentrations) and at a KCl/HCl (3:1) || KCl (same concentration as HCl) contact zone. [In this illustration it is assumed that H^+ ions migrate twice as fast as K^+ ions (in fact ~ 5x). This is represented in the figure by showing all of the H^+ ions at the interface migrating in the direction of the concentration gradient, but only every second K^+ ion.]

diffusion is detectable. This means that more positive charges are migrating into the KCl solution than into the HCl solution. This results in a positive charging of the KCl side of the contact zone relative to the HCl side. This charging continues until the opposing electrostatic field slows down any further migration of H^+ ions from HCl solution into the KCl solution, and a state of equilibrium has been reached.

2.3.2 Calculation

Unfortunately, the magnitude of the liquid junction potential resulting from charge separation at the point of contact of two different electrolyte solutions can seldom be theoretically calculated (only when either the cation or anion is the same, or when the concentration of the ions is the same in both phases). In these cases the exact compositions of the adjacent solutions must be known, a requirement which is seldom met in practice.

In the derivation of the formula for calculating the liquid junction potential, the electric work done in separating the charges is set equal to the work of diffusion; that is, the change in chemical potential arising from the diffusion of the ions. Only after making certain *approximations* can one arrive at the so-called Henderson solution [56] of the Nernst-Planck equation [57]:

$$\epsilon_{\mathrm{lj}} \approx -\frac{R \cdot T}{F} \sum_{\mathrm{i}} \frac{t_{\mathrm{i}}}{z_{\mathrm{i}}} \ln \frac{a_{\mathrm{i1}}}{a_{\mathrm{i2}}} \tag{26}$$

where t_{i} = transport number of ion of type i
$a_{\mathrm{i1,2}}$ = activity of ion i on side 1 and 2, respectively

For purely monovalent electrolytes, for example KCl, the most commonly used in salt bridges, this becomes:

$$-\epsilon_{\mathrm{lj}} \approx \frac{R \cdot T}{F} (t_{+} - t_{-}) \ln \frac{a_{\mathrm{i1}}}{a_{\mathrm{i2}}} \tag{27}$$

where t_{+} = transport number of the cation
t_{-} = transport number of the anion.

The liquid junction potential calculated with this equation should not be used for very accurate applications. The conditions assumed in its derivation are never all fulfilled in practice. The main assumption, that the participating ions migrate exclusively according to the concentration gradients so that pure diffusion results, is seldom valid. This is true especially for the common reference electrode constructions in which convection of the electrolyte arises (this is desired for other reasons). In addition, in order to calculate the individual ion activities one needs the individual activity coefficients, and these are not accessible through experimental measurements (as shown in the Appendix). *In practice, the analytical applications of EMF measure-*

ments do not depend on the absolute magnitude of the liquid junction potential, but only on its constancy. Its magnitude can be cancelled out by the necessary calibration procedures. Comparison of the liquid junction potentials calculated with Eq. (27) with those found experimentally shows discrepancies of between 1 and 2mV relative to an absolute value of about –30mV for contact zones of the type: HCl || alkali chloride solutions at equal concentrations [58].

In the literature electrode potentials which are determined using an electrochemical cell with liquid junction are often followed by the notation $+\epsilon_j$.

2.3.3 Salt Bridge Electrolytes

A glance at Eq. (27) shows what is to be done in order to keep the liquid junction potential small. Aside from the trivial case in which $a_{i1} = a_{i2}$ (same electrolyte and same concentration on both sides), its magnitude depends above all on the factor in front of the logarithm; on the difference in transport numbers of the cation and anion. One can attempt to minimize its value by choosing cations and anions with similar mobilities for the salt bridge electrolyte. Potassium chloride ($t_{K^+} - 0.49$, $t_{Cl^-} = 0.51$), for example, is a suitable salt. KNO_3, NH_4NO_3, RbCl, etc. also exhibit equally favorable salt bridge electrolyte properties.

The concentration appropriate for the salt bridge electrolyte depends on the properties of the corresponding sample solution. If the sample solution contains large and variable concentrations of ions with large migration velocities (strongly acidic or basic media), then a high concentration of salt bridge electrolyte is required. It can be shown (compare with Fig. 16) that in this case the larger transport number of H^+ and OH^- ions can be compensated for by the high concentration of salt bridge ions of nearly equal mobilities, since at high enough concentrations these ions carry most of the current. Nonetheless, the liquid junction potential can never be completely eliminated in this way. For measurements made in strongly acidic or basic media one must always reckon with liquid junction potentials much greater than 10mV and their associated variations [17, 59–63].

Although a saturated solution of KCl is most often employed as a salt bridge electrolyte and also as a reference electrode electrolyte, there are practical reasons for advising against the use of saturated solutions. With saturated salt bridge electrolyte solutions there is always the danger that with cooling or with long storage periods (and evaporation) the salt will crystallize out somewhere in the electrolyte bridge, strongly interfering with the current flow. This frequently occurs in practice. When this happens one will either see actual crystals in the immediate vicinity of the contact zone, or will notice irreproducible deflections of the detecting instrument, indicative of some obstruction of current flow in the capillary regions connecting the two solutions. *Far more than half of the failures of EMF measurements in cells with liquid junction can be traced back to some disturbance in the contact zone.*

Getting salts which have crystallized in the capillary canals back into solution is very time-consuming and requires exchanging the saturated solution for a more dilute one in order to get the salt completely dissolved again. In many cases the electrolyte bridge and with it often the entire reference electrode must be discarded. In addition to crystallization of the main electrolyte salt, very dilute sample solutions can also lead to unexpected precipitates. For example, AgCl may precipitate at the contact zone, since a concentrated KCl solution allows greater dissolution than a dilute one. This electrode salt will immediately precipitate out once the electrolyte concentration is lowered through mixing with the dilute sample solution. This occurs right in the diaphragm capillaries when working with very dilute sample solutions. Clogging of the diaphragm in this manner is not always immediately recognized as deflections of the recording instrument, since AgCl exhibits a finite ionic conductivity. Such clogging will result in an unstable liquid junction potential, which manifests itself as too large a standard deviation in the final results. The AgCl precipitate can frequently be dissolved by dipping the tip of the salt bridge in concentrated ammonia solution; it is better, however, to prevent its formation in the first place by using a less concentrated KCl solution. It is advantageous to use 0.1 to 1 *M* KCl solutions when making measurements with very dilute (low ionic strength) sample solutions.

Saturated salt bridge solutions have still further drawbacks. If a reference element is immersed in solution and it can undergo a reversible electrode reaction with an electrolyte ion (Cl^- ion), then it provides a stable equilibrium Galvani potential, the magnitude of which depends on the activity of the corresponding reference ion according to the Nernst equation. If the solution is saturated, as is the case with most commercially available reference electrodes, then this activity is often a sensitive function of the temperature of the solution. The stronger the solubility of the salt bridge electrolyte depends on temperature, the greater the change with temperature of equilibrium Galvani potential at the reference element/solution interface. In this way relatively small temperature changes which are difficult to eliminate between individual measurements can cause variations in the reference potential, and thus errors in the results. This is especially true if a symmetric cell arrangement is not used. In a *symmetric cell* an identical reference electrode exists on the other side of the ion-selective membrane, i.e., inside the ion-selective electrode. Since the potential variations caused by solubility changes are equal on both sides of the membrane, they compensate for one another. Only a very few ion-selective electrodes contain an element immersed in a *saturated* electrolyte solution; indeed many have a direct solid state contact. Hence one should not routinely work with saturated solutions in the outer reference electrode. *For all these reasons it is advisable to use a concentrated, yet not completely saturated solution as a salt bridge electrolyte.* The concentration should be far enough from the saturation point, that at the lowest anticipated temperature still a long time period is required before any crystals are formed through evaporation (for KCl between 1 and 3 M).

2.3.4 Construction of the Salt Bridge Electrolyte/Sample Solution Contact Zone

Such drastic simplifying assumptions were made in the derivation of Eq. (26) for the liquid junction potential that accurate predictions of its exact magnitude are not possible. In addition, most of the commercially available reference electrodes with salt bridges do not have interfaces which are locally stable enough to conform to the assumptions in this derivation. These electrodes often show a streaming of the concentrated salt bridge electrolyte (mostly KCl solution) into the test solution (desired for other reasons). Experience shows that the liquid junction potentials vary with the construction of the mixing zone.

The lowest flow rates (from approximately 0.005 to 0.1 ml/24hr) are exhibited by salt bridges with asbestos or linen fibers fused in glass. Ungreased glass or plastic ground sleeve diaphragms coated with electrolyte have the largest (about 1 to 50 ml/24hr). The large number of currently available materials with vastly different porosities (glass or ceramic frits, inert sintered materials, etc.) make possible salt bridge connections with almost any desired flow rate (see Fig. 17).

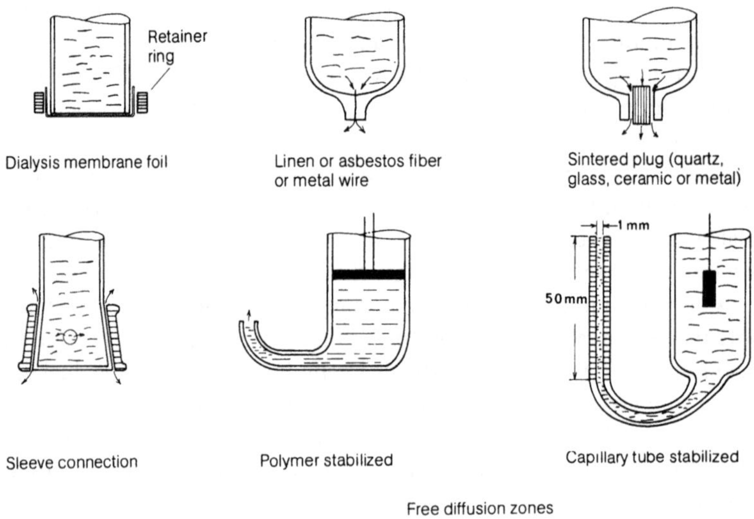

Fig. 17. Possible salt bridge/sample solution contacts

In the medicinal and biological fields precipitation of proteins in the connecting capillaries, and thus clogging of the salt bridge, can easily occur in the ways previously described (concentrated solution of a strong electrolyte). Recently cellophane dialysis membrane foils have proven to be the best auxilliary aids for establishing locally stable diffusion interfaces [64]. Free, large surface area diffusion zones stabilized with agar-agar which have a heavier salt bridge electrolyte beneath the test solution also yield excellent results in medicinal applications. In order to obtain the required stable contact zone in this case, a fresh portion of the polymer-stabilized electrolyte is used for each measurement. This is accomplished by forcing a few millimeters of the agar-agar stabilized KCl solution out after each measurement with the help of a piston, and washing it away by rinsing out the cell containing the sample solution (Fig. 18). With a concentrated KCl solution (without agar-agar) contacting the test solution below it only via a precisely bored 5 cm long, 1 mm diameter capillary tube, a liquid junction potential stability of better than ± 0.04mV over a number of hours was obtained [65].

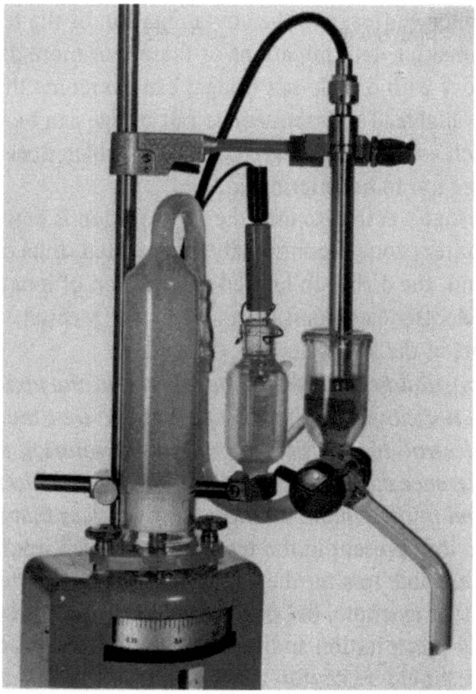

Fig. 18. Precision measurement cell (using a polymer stabilized salt bridge and a free diffusion zone – Metrohm)

Systems with the lowest electrolyte flow rates are ideal from an analytical standpoint because they contaminate the sample solutions least. Unfortunately, practice has shown that such systems show the least stable liquid junction potentials. Hence they are not to be recommended for precise measurements with ion-selective electrodes. Almost all the usual pH reference electrodes fall into this category. In this connection it must be pointed out that difficulties arise with asbestos fiber bridges when measuring in solutions buffered with 2-amino-2-hydroxymethyl -1,3-propanediol; platinum wires fused in glass are also not to be recommended for use in strong redox media. *It has been shown that under routine conditions the most accurate measurements for cells with liquid junction can be obtained by using ground sleeve diaphragms* [66, 17]. Schwabe et al. were able to show that even with extremely different sample solutions, (large H^+ or OH^- ion concentration differences) the liquid junction potential remained constant to within ±0.2mV when using sleeve salt bridge connections [17]. In addition, sleeve diaphragms are ideally suited for highly contaminated (river water, soil suspensions, etc.) or very viscous (salves) samples, because the contact zone is easily cleaned by removing the outer ground sleeve. The electrical resistance of such connections is small, and with sample solutions of too low electrolyte concentration the large electrolyte flow quickly provides the conductivity needed for the measurement.

Disadvantages of sleeve diaphragms are the large consumption of electrolyte which renders them less maintenance-free, and the contamination of the test solution by salt-bridge electrolyte ions, making determinations of these ions more difficult. Suitable sleeve reference electrodes with double salt bridges can overcome this last difficulty and are commercially available. The intermediate salt bridge can be filled with a second inert electrolyte such as NH_4NO_3, $Mg(NO_3)_2$, etc., which does not contaminate the test solution with the ion to be determined.

A flow of the salt bridge electrolyte into the test solution is desired for a number of reasons: First, the contact zone is continually regenerated, thus retaining its characteristic features; second, the diffusion limited penetration of ions from the test solution into the reference electrode compartment is reduced (precisely the reason for using cells with transport in the first place).

In view of this last possibility, care should be taken that the upper fill-hole for the salt bridge electrolyte is not closed with an air-tight seal, so that the hydrostatic pressure of the column of electrolyte is sufficient to allow the solution to flow out. For the same reason, the reference electrode should not be immersed in the test solution so far that the liquid level in the sample solution is higher than that in the reference electrode. There may be ions present in the test solution which upon penetration into the reference electrode are able to alter the equilibrium Galvani potential of the reference electrode half-cell (for example, Br^- or I^- with an Ag/AgCl internal element).

In order to avoid any precipitation and the associated problems discussed above, the tip of the salt bridge should be kept in a *pure, somewhat diluted solution of the salt bridge electrolyte* when not in use.

2.3.5 Cells Without Liquid Junction

In view of all the difficulties associated with liquid junction potentials, it is desirable to construct cells without liquid junctions. In practice this is possible more often than one might think. All that is needed is for the sample solution to contain an ion with constant activity which does not interfere with the ion-selective indicator electrode used, and which can itself be specifically sensed with a second ion-selective electrode. This second ion-selective electrode can then function as a reference electrode, since it fulfills the primary requirement of a good reference electrode, i.e. it has a constant equilibrium Galvani potential at the interface: electrode/solution. With such a set-up problems of a technical nature arise only if the resistance of the ion selective electrode connection with the reference electrode jack is too large. In this case an ion-meter with a high-ohmic differential input must be used. Such devices are already sold commercially (see Fig. 43).

If the test solution does not contain any detectable ion in a constant, stable amount, as is the case in most analytical work, then a suitable compound which does not interfere with the indicator electrode can often be added before the EMF measurement. The amount of such an ion-pair (an ion can only be added as a cation-anion ion-pair) needed to produce a constant equilibrium Galvani potential at the second electrode can lie between 10^{-2} and 10^{-5} M, if this ion is originally present in the test solution only in traces. For cases where the individual sample solution contains variable amount of this ion, solutions which are saturated in this reference ion can be used, provided that both electrodes are still usable in such concentrated solutions. A calculation of the error involved if the solution is not saturated with the reference ion is advisable. As an example, if the *variation* in concentration of the chosen reference ion in the sample solutions are about 0.01 mole/l and an amount of reference ion corresponding to a final concentration of 1 mole/l is added, then an error of 1% results. For reference and measured ions of the valence, this error in reference ion concentration is fully reflected in the final results.

In many cases where ion-selective electrodes are employed, the test solutions must be buffered; for example if H^+ ions are indicated by the measuring electrode (pNa, pK, pCa, . . .), or if H^+ ions combine with the ion to be determined leading to undissociated or complexed species ($F^- + H^+ \rightleftharpoons HF + F^- \rightleftharpoons HF_2^- + F^- \rightleftharpoons$ . . .), or if H^+ ions can disturb an equilibrium of the measured ion ($S^{2-} + H^+ \rightleftharpoons HS^-$; $NH_3 + H^+ \rightleftharpoons NH_4^+$). In many of these cases a pH glass electrode can serve as reference electrode. In another case a sodium-specific glass electrode has been successfully used as reference electrode for fluoride ion determinations [58]. The reverse case can also be imagined, or a fluoride-specific electrode for use in other halogen determinations, etc. Preparing solutions with the appropriate, constant reference ion activity need not be an additional operation with this type of cell arrangement. Often, for reasons already mentioned, a buffer or interfering-ion-suppressing agent must be added. If all these additional reagents are combined in a single conditioning solution (i.e. TISAB, ADE, . . .), then the sample solution can be measured after only one manipulation.

2.4 Secondary Reference Electrodes

It has already been pointed out that under suitable conditions the normal hydrogen electrode provides the most stable equilibrium Galvani potential, and is therefore to be recommended for very precise measurements. In such applications the hydrogen electrode is not always immersed in an acid solution with $a_{H^+} = 1$, since this relatively high H^+ ion concentration leads to a relatively large liquid junction potential. This cannot be fully compensated for even with the use of a suitable salt bridge (a liquid junction potential of about 5mV can be calculated for the system: 0.1 M HCl || sat. KCl [67, 68]). If the hydrogen electrode is dipped in one of the standard buffer solutions recommended by NBS and the DIN 19266, the EMF of the resulting cell can be easily converted to that for pH = 0 with the use of Eq. (24). Thus the standard electrode potential ϵ^0 of the second electrode present in the cell can be obtained.

Since the handling of the normal hydrogen electrode is too bothersome for many applications, the less complicated electrodes of the second kind are usually employed as secondary standards in routine measurements. In these cases the cell EMF differs by a constant additive term from the EMF which would be shown by the corresponding measuring electrode in combination with a hydrogen electrode. Thus one can easily convert this measured potential to a potential relative to the standard hydrogen electrode potential.

2.4.1 Silver/Silver Chloride

A silver wire covered with a layer of silver chloride responds to the chloride activity of the contiguous solution according to:

$$\Delta\phi_{eq} = \Delta\phi'_{eq} - \frac{R \cdot T}{F} \ln a_{Cl^-}. \tag{28}$$

If the sample solution has a constant chloride ion activity, then this element can be dipped directly into the test solution to construct a cell without liquid junction. In all other cases (strong oxidizing or reducing media; presence of AgCl complexing agents such as CN^-, SCN^-; presence of species which form more insoluble compounds with Ag^+ such as S^{2-}, Br^-, I^-; etc.) an approximately 0.1 to 3 M KCl solution is inserted between the internal Ag/AgCl element and the sample solution by means of a salt bridge. *The solubility of AgCl in concentrated KCl solution is considerably larger (~ 10^{-2} M/l) than that in pure water. For this reason the KCl solution used should be saturated with AgCl by adding enough AgCl so that a small amount remains as a solid at the bottom of the solution. Otherwise, after a few fillings of the salt bridge, and especially at higher temperatures, the AgCl layer of the element will dissolve, leading to irreproducible results.* A complete Ag/AgCl reference electrode with liquid junction can be symbolized by:

$$\text{Ag} \mid \text{AgCl} \mid \text{KCl (0.1 to 3 M)} \parallel \ldots$$

2.4.1.1 Preparation

There are many ways to bring an AgCl layer in contact with a silver support. The easiest way to accomplish this is to electrolytically deposit AgCl on a silver wire anode in an approximately 0.1 M chloride solution. One can observe this deposition by the brown coloration of the electrode surface. If care is taken to work with pure solutions, Ag/AgCl elements can be produced which agree with one another to within ±3mV. Silver-coated copper, gold or platinum wires can also be used, of course. According to another procedure [69], a freshly precipitated, well washed aqueous Ag_2O paste is placed inside a platinum wire spiral, converted into a porous silver bar by heating at high temperature, and then anodically covered with a chloride layer as described above. With this technique agreement between the individual Ag/AgCl elements of better than ±0.02mV can be attained. The required AgCl layer can also be obtained by dipping a silver wire in an AgCl melt [70]. Deviations among the potentials of various electrodes produced in this way can reach as high as 0.2mV [71, 72].

2.4.1.2 Characteristics

Next to the hydrogen electrode, the Ag/AgCl half-cell is the most reproducible half-cell. Its only drawback is that in acidic electrolyte solutions it can react with dissolved oxygen according to

$$2Ag + 2H^+ + 2Cl^- + \tfrac{1}{2}\,O_2 \rightleftharpoons 2AgCl + H_2O, \qquad (R8)$$

which can lead to a shift in the equilibrium Galvani potential as large as 1.5mV. Only if one wants to attain absolute accuracies better than 0.3mV when using neutral KCl solutions, is it advisable to first bubble the filling solution with an oxygen-free inert gas [73].

Due to the relatively large solubility of AgCl in 3 M KCl solution, Ag^+ ions pass through the salt bridge diaphragm into the sample solution along with the K^+ and Cl^- ions. Thus one should not try to determine Ag^+ ions, or obviously K^+ or Cl^- ions, using such a reference electrode. In addition, the utmost care is required when working with liquid junctions which are difficult to clean to be sure that the sample solution does not contain any ions which form other insoluble salts with Ag^+ ions, such as Br^-, I^-, S^{2-}, etc. These precipitates deposit in the canals of the diaphragm and block the necessary electrolytic contact between the test solution and the reference electrode solution. With the exception of the sleeve diaphragms the cleaning of such clogged contact zones is nearly impossible, and due to the electrode construction, the entire reference electrode must be thrown away. *The best way to avoid these problems when working with such solutions, or when measuring Ag^+, K^+ or Cl^- ions, is to use an Ag/AgCl reference electrode in conjunction with a double salt bridge. Such electrodes are commercially available* (Fig. 19). Any inert electrolyte with about the same transport number for cation and anion which does not interfere with the indicator electrode or react with ions present in the sample solution can be chosen as a second, outer salt bridge electrolyte (for example KNO_3, NH_4NO_3, NH_4Cl, $Mg(NO_3)_2$, etc.). Frequent changing of this electrolyte is advisable to be certain that no interfering ions become concentrated there, since they would eventually find their way into the sample solution.

The potential of the Ag/AgCl reference electrode is temperature dependent. At 25°C with 3.5 M KCl it has a value of +200 ±5mV. This amount is to be added to each measured EMF to refer it to the normal hydrogen electrode. If the potential of the Ag/AgCl reference electrode is determined at various temperatures between 0 and 95°C by measuring the EMF relative to a hydrogen electrode, whenever the electrode is warmed or cooled back to room temperature, the electrode potential returns to within ±2mV of its original value. This *temperature hysteresis* effect is thus very small with the Ag/AgCl reference electrode. For this reason it is recommended for all applications in which the temperature cannot always be held constant or lies above 80°C.

An Ag/AgCl reference electrode can easily be tested to see if it is functioning properly by checking it against a new Ag/AgCl electrode prepared according to one of the above procedures. The potential difference between the electrode to be tested and the new Ag/AgCl element, when both are immersed in filling solution can be measured with a pH meter, and should show a stable value of between ±5mV. If the

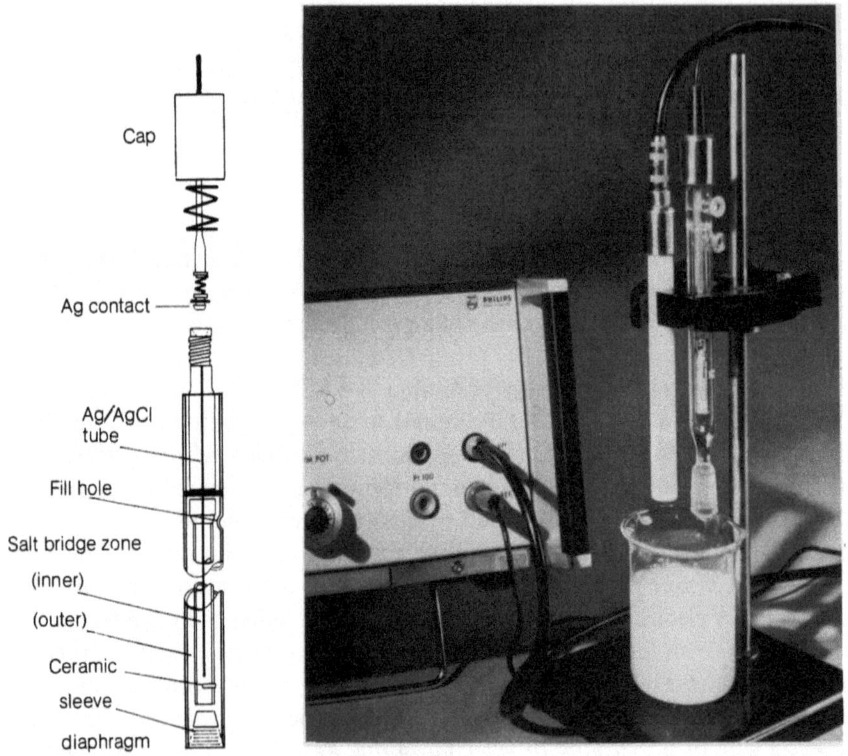

A B

Fig. 19. A: Schematic diagram of a reference electrode with double salt bridge and sleeve diaphragms. A construction of Orion made of plastic, allowing measurements in acidic HF solutions. **B:** Complete measuring cell consisting of an ion-selective indicating electrode and a reference electrode with double salt bridge and sleeve diaphragm (Philips)

meter indicates irreproducible potentials or a large deflection, then the electrolyte contact zone is probably clogged. If this plug is only precipitated KCl, then the electrode can be returned to its original condition by warming the tip for a prolonged period of time in 0.1 M KCl solution. This redissolution can usually be hastened by slipping a vacuum tube over the clogged electrode tip and applying a water aspirator. If this test yields a stable potential greater than ±5mV, then the inner filling solution is probably contaminated. Traces of bromide or iodide can easily effect large potential shifts!

2.4.2 Thallium Amalgam/Thallium (I) Chloride (Thalamid®)

In its mode of operation this element corresponds exactly to the Ag/AgCl element, except that in this case thallium metal is unstable in aqueous electrolytes so its amalgamated form is employed. Just as with the more well-known alkali amalgam electrodes, a thallium amalgam electrode responds reversibly and according to the Nernst equation to the Tl^+ ion activity in the contiguous solution. If the Tl^+ ion activity is controlled by some anion to yield a sparingly soluble precipitate, as is usual for electrodes of the second kind, then the electrode will respond to the activity of this anion, just as the corresponding Ag/AgCl element responds to the Cl^- ion activity. Thus, if TlCl is brought in contact with thallium amalgam, the electrode potential reflects the chloride ion activity of the adjacent solution according to:

$$\Delta\phi_{eq} = \Delta\phi'_{eq} - \frac{R \cdot T}{F} \ln a_{Cl^-} \; . \qquad (29)$$

Fricke [74] was the first to propose that this reversible and wide-ranging unpolarizable element be used as a reference electrode in conjunction with a KCl salt bridge. The Thalamid® electrode of Jenaer Glaswerke Schott & Gen, Mainz, [75] can be symbolized by:

(Pt) Hg, Tl | TlCl | KCl (3 M) || ...

2.4.2.1 Characteristics

The reason that the Thalamid® reference electrode is introduced here before the much more well-known Calomel reference electrode stems from its far superior temperature stability. It shows no temperature hysteresis up to 135°C, and in this respect is even superior to the Ag/AgCl reference electrode. As with the Ag/AgCl electrode, the reaction

$$2Tl + 2H^+ + 2Cl^- + \tfrac{1}{2}O_2 \rightleftharpoons 2TlCl + H_2O \qquad (R9)$$

requires that, for extremely precise absolute measurements, the filling solution should be deoxygenated with an oxygen-free inert gas prior to use. The solubility of TlCl in 3 M KCl is 10^{-5} M/l, much lower than that of AgCl. Nonetheless, the appropriate precautions should be observed, and interfering substances in higher concentrations which are the same as for the Ag/AgCl electrode, should be kept away from direct contact with the reference electrolyte via an additional salt bridge. The standard potential of this reference electrode is of course also temperature dependent. At 25°C with a 3.5 M KCl electrolyte it has a value of −575 ±5mV. Because this value is so large relative to the other usual reference electrodes (Ag/AgCl and Calomel), replacing one of these reference electrodes with the Thalamid® reference electrode can lead to some problems of a technical nature. Since Ag/AgCl is used as the internal element of many ion-selective electrodes, the potential which results from this unsymmetrical cell is about 800mV. This large asymmetry potential cannot always be compensated for by the asymmetry or calibration control of the pH meter used (the usual potential span of ±200mV is too small).

Proper electrode function can be tested with a freshly prepared Ag/AgCl element as described in the testing of the Ag/AgCl reference electrode. In this case a stable potential of ±775 ±5mV, depending on the input connections, should be indicated on the pH meter.

2.4.3 Mercury/Mercury (I) Chloride (Calomel)

The well-known Calomel reference electrode is by far the most widely used reference electrode. It is symbolically written:

$$(Pt)\ Hg \mid Hg_2Cl_2 \mid KCl\ (3M) \parallel \ldots$$

It need not be explained in greater detail here. It, unfortunately, exhibits so large a temperature hysteresis effect due to the temperature-sensitive disproportionation

$$Hg_2Cl_2 \rightleftharpoons Hg + HgCl_2, \qquad (R10)$$

that it cannot be used at temperatures above 75°C. Systematic studies [59] have shown that it should have only limited use as a reference electrode because its equilibrium Galvani potential is subject to large fluctuations. The Calomel electrode should be replaced by the hardly more expensive Ag/AgCl electrode, or in cases where the Ag^+ ion interferes, with the Thalamid® electrode. For a comprehensive survey of the various reference electrode systems the reader is referred to the book by D.H.G. Ives and G.J. Janz [53].

3 Ion-Selective Electrodes

3.1 General Construction Principles

Depending on the ion-sensitive material used, one can following recent IUPAC recommendations distinguish between

crystalline and *noncrystalline electrodes.* The first can be further divided into *homogeneous* (e.g., single crystal, pressed or melted pellet) and *heterogeneous* (e.g., AgCl in PVC, silicone rubber) *"solid" electrodes.* The noncrystalline electrodes can be subdivided into *porous supported* (e.g., Millipore filter, glass frit) and *nonporous supported* (e.g., PVC) *liquid ion exchanger* and *neutral carrier electrodes.*

Due to the relatively poor electrical conductivity of these electrode materials, the thinnest possible layers are used, and they are thus called membrane electrodes. The word membrane, however, should not lead to the erroneous assumption that one is dealing with membrane thicknesses of biological dimensions. Depending on the resistance of the particular material, the membrane thickness can vary considerably: 0.1 mm for glass, 1 to 5 mm for organic materials, and greater than 3 mm for single crystal and pressed pellet electrodes.

The problem of how to transmit, unaltered, the potential of the ion-sensitive phase still remains. It is not always possible to simply imbed one end of the instrument lead in the material which carries out the specific ion exchange with the test solution. In this case an additional phase boundary, with its own potential, must be introduced into the measuring circuit. In order for a constant equilibrium Galvani potential to arise at this interface, a reversible reaction is required. This contact point must also behave like an unpolarizable electrode, a requirement which is not as easily fulfilled. One already has, however, a reversible and unpolarizable electrode reaction on the measuring side of the membrane (otherwise the electrode response would not follow the Nernst equation). The obvious solution, then, is to allow the same electrode reaction to occur on the inner side of the membrane, only this time with a constant activity of the corresponding measured ion.

Thus the problem of how to further transmit the potential from the interior of the ion-selective phase is solved, insofar as one can be content with an additional, stable Galvani potential at this new interface. Meanwhile, the same problem exists with this new inner solution phase as that which we encountered in the measurement of a single potential difference: namely, how does the potential of this inner solution get to the input terminal of the voltmeter? Once again there is no choice other than to be content with the construction of an additional interface and the inclusion of yet another potential difference. Again, care must be taken that the Galvani potential here is also stable. This can be accomplished, as we have seen before with a reference element. We must only make sure that the activity of the ions participating in the reactions at both interfaces of the inner solution remains constant. If the same reference element is used outside (measuring side) and inside (reference side) then a symmetric measuring circuit results. Due to its compensating effects, such a circuit is preferred. Fig. 20 illustrates the construction of ion-selective electrodes.

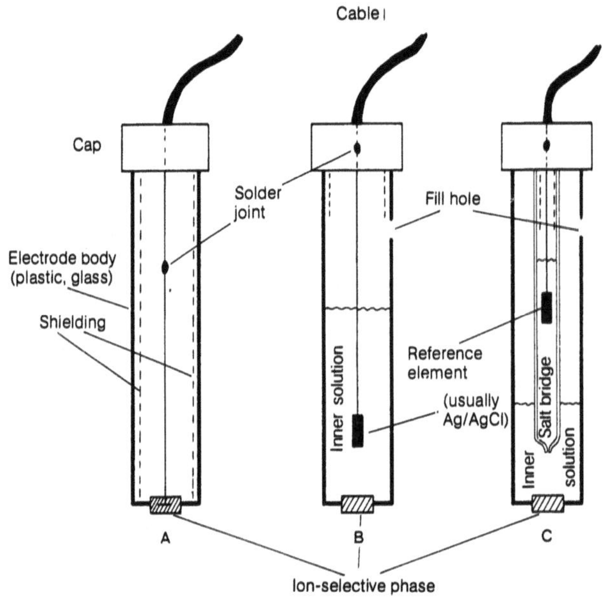

Fig. 20 A–C. Possible constructions of ion-selective electrodes with solid active phases. A: Direct contact of the metallic conductor with the active phase. B: Contact through an inner solution and a reversible shunt element. C: Contact through an inner solution and an ordinary reference electrode with salt bridge

In a few cases (for example with some solid-state membrane electrodes) it is possible to keep unpolarizable the inner contact point to the ion meter, where the crossover from ion conduction to electron conduction occurs (reversible redox reaction). Such solid-state contacts eliminate the need for inner solutions with constant ion activities, and are thus practically maintenance-free. To be sure, with constructions of this type one must keep in mind that there is no longer a compensating inner reference electrode. The temperature dependence of a cell containing such an electrode may be entirely different from that of a symmetric cell; in addition, the polarizability can be larger. There is much freedom possible in the design of individual electrode shapes (see Fig. 21).

In the descriptions of the individual electrodes which follow, their most important properties and their proper handling are discussed, and indications of sample preparation methods are given (insofar as precautions are necessary to insure interference-free determinations). The measuring procedure is somewhat the same with all electrodes and is described later in Chap. 4 and 5. Keep in mind that the division

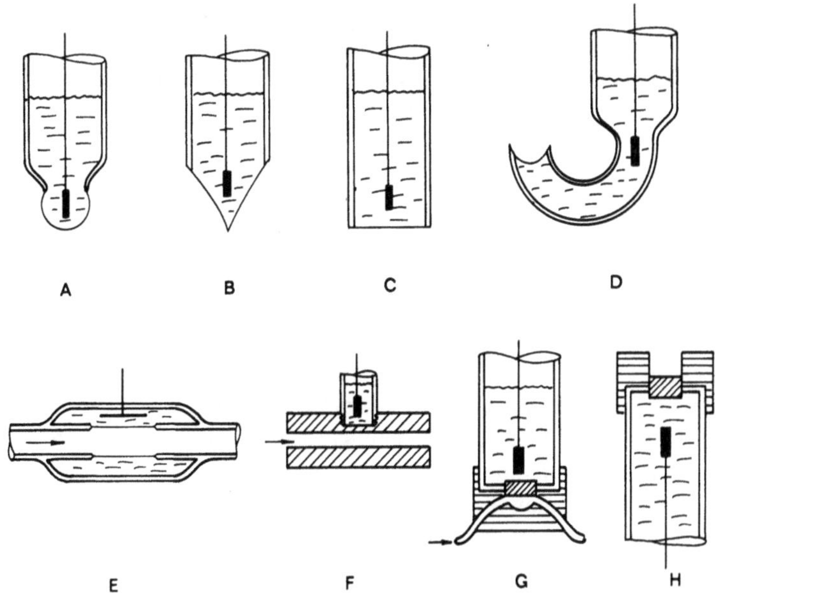

Fig. 21 A–H. Possible indicator electrode construction designs. **A:** Ball form (usually with glass membrane electrodes), **B:** Pointed form (usually with glass membrane electrodes), **C:** Flat form (glass membrane and solid-state electrodes), **D:** Convex form (glass membrane electrode), **E:** Flow-through design (glass membrane electrode), **F:** Flow-through design (solid-state electrodes), **G:** Flow-through design (liquid membrane electrodes), **H:** Microliter arrangement (solid-state electrodes)

into categories was oriented toward differences in construction, so that some properties and operations could be discussed in general. The sequence derives from the extent to which the various types are currently used for didactic reasons, and does not represent a value judgment.

3.2 Solid-State Membrane Electrodes

3.2.1 Glass Membrane Electrodes For Li^+, Na^+, K^+, Rb^+, Cs^+, NH_4^+, NR_4^+, Ag^+, Tl^+ Ions

3.2.1.1 Construction

We will pass over here the classic pH glass electrode. Those who wish to learn more about the topic of electrochemical pH measurements are referred to the monographs of M. Dole [5], L. Kratz [6], K. Schwabe [17] and G. Eisenman [7]. The last of these already describes the transition to ion-selective glass membrane electrodes.

The investigations of Eisenman et al. [16] concerning the alkali error of pH glass electrodes led to the development of glasses of the aluminum silicate series, which in addition to a sensitivity for hydrogen ions show a high specificity for sodium ions. Typical sodium-selective glass compositions are, for example:

NAS_{11-18} (11% Na_2O, 18% Al_2O_3, remainder SiO_2) [16]

$NAS_{10.6-10}$ (10.6% Na_2O, 10% Al_2O_3, remainder SiO_2) [19]

$LAS_{26.2-12.4}$ (26.2% Li_2O, 12.4% Al_2O_3, remainder SiO_2) [Beckman number 39278].

With sodium-selective glasses the preference for Na^+ ions over other monovalent ions (with the exceptions of H^+ and Ag^+ ions) is so marked, that electrodes made of these glasses can be used without problems in almost all cases.

The difficulties with building a correspondingly specific potassium electrode have not yet been successfully overcome. Glasses which respond to K^+ ions also respond equally well to all the other monovalent ions listed in the section heading. Electrodes of this type are referred to as cation-sensitive electrodes. These glasses consistently have a lower Al_2O_3 content, for example:

NAS_{27-4} (27% Na_2O, 4% Al_2O_3, remainder SiO_2) [19]

KAS_{20-5} (20% K_2O, 5% Al_2O_3, remainder SiO_2) [19]

$KABS_{20-5-9}$ (20% K_2O, 5% Al_2O_3, 9% B_2O_3, remainder SiO_2) [76].

pNa electrodes are characterized by selectivity coefficients $k^{pot}_{Na\text{-}K}$ of between 10^{-2} and 10^{-3}, thus showing a high specificity for Na^+ ions relative to K^+ ions. On the other hand, the corresponding values of $k^{pot}_{K\text{-}Na}$ for the potassium-sensitive electrodes are between 0.1 and 0.2. This is not sufficient to allow the determination of potassium in the presence of varying amounts of sodium. Ion-selective glass membrane electrodes have been offered for sale by almost all pH electrode manufacturers. As Table 1 shows, both types of electrodes exhibit strong responses to both H^+ and Ag^+ ions.

Both types of ion-selective glass membrane electrodes available to date are constructed in the same way as normal pH electrodes. As a rule, the glass electrode consists of a hemisphere blown out of the special glass, which is fused to a shaft of poorly conducting apparatus-glass in order to avoid conduction along the surface of the glass and current leaks. An exception is the sodium electrode built by Orion, in which a disk of sodium-selective glass is cemented into a plastic electrode body. The glass disk is cut from a special heat-treated glass rod, a manufacturing technique which guarantees invariant properties.

An inner solution with a constant activity of the respective measured ion, sodium or potassium, is contained in the tip of the electrode. To deliver the potential further, a reference element, usually Ag/AgCl, dips into the inner solution forming a reference cell without liquid junction. The inner solution must contain a constant chloride ion activity so that the equilibrium Galvani potential of this reference electrode remains constant. There are also now micro-flow-cells made with sodium-selective glass (Fig. 22) which require only about 50 μl for a determination.

Table 1. Glass membrane electrodes

Electrode Type	Active Phase	Potential-determining Ions	Working range [M]	Selectivity coefficient[1] $k^{pot}_{M\text{-}I}$	Recommended pH-range	Temperature range [°C]	Electrical resistance at 25°C [MΩ]	Recommended reference electrode	Manufacturer see also Appendix A.8
pNa	NAS_{11-18} $LAS_{26.2-12,4}$ etc.	$Ag^+ > H^+ > Na^+ > K^+$	1 to 10^{-8}	$Ag^+ \sim 500$; $H^+ \sim 10^3$; $K^+ \sim 10^{-3}$; $Li^+ \sim 10^{-3}$; $Cs^+ \sim 10^{-3}$; $Tl^+ \sim 2 \cdot 10^{-3}$; $Rb^+ \sim 3 \cdot 10^{-5}$; $NH_4^+ \sim 3 \cdot 10^{-5}$	7 to 10 ca. 4 pH units above pNa-value	0 to 100	~ > 100	Ag/AgCl with double salt bridge (1 M NH_4NO_3)	Beckman, Corning, EIL, Ingold, Metrohm, Orion, Philips, Polymetron, Radelkis Radiometer, Schott & Gen Tacussel
pCation	NAS_{27-4} $KABS_{20-5-9}$ etc.	$H^+ > Ag^+ > K^+$; $NH_4^+ > Na^+ > Li^+$ Rb^+, Cs^+, Tl^+	1 to $5 \cdot 10^{-6}$	$Na^+ \sim 0.1$; $NH_4^+ \sim 0.3$; $Rb^+ \sim 0.5$; $Li^+ \sim 0.05$; $Cs^+ \sim 0.03$ (normalized to: $K^+ = 1$)	7 to 13 (for pK) 4 to 10 (for pAg) ca. 2 pH above pK-value	0 to 100	~ > 100	Ag/AgCl with double salt bridge (1 M Li-trichloroacetate)	Beckman, Corning, EIL, Ingold, Philips, Tacussel

[1] According to the manufacturer

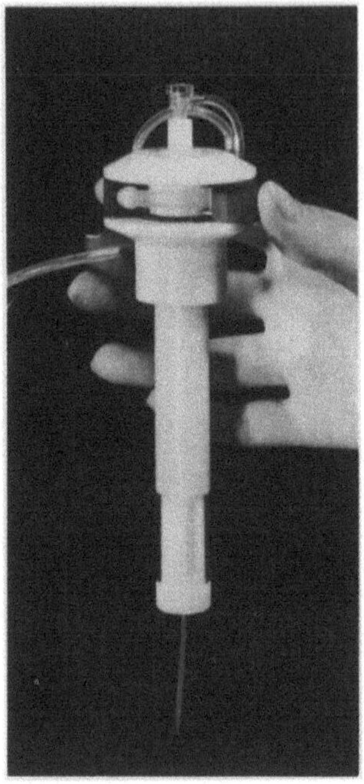

Fig. 22. Microliter flow-through cell made of sodium-selective glass (Corning)

3.2.1.2 Characteristics

An ion-selective glass membrane electrode dipping in a solution of the corresponding measured ion, and complemented with a suitable reference electrode forms a cell (with or without liquid junction) which will exhibit an EMF according to the equation

$$E = E' + \frac{R \cdot T}{F} \ln \left(a_M + \sum_I k_{M\text{-}I}^{pot}\, a_I^{1/z_I}\right) \tag{30}$$

where E' = EMF of the measuring circuit with a standard solution; E' depends, among other things, on the activity of the inner solution and type of reference electrode employed
a_M = activity of the measured ion
a_I = activity of the interfering ion
z_I = valence of the interfering ion
$k_{M\text{-}I}^{pot}$ = selectivity coefficient: measured ion-interfering ion.

Since sodium-selective glass membranes are very selective for Na^+ ions, the product of the selectivity coefficient and the interfering ion activity can be neglected in many cases, leaving, at 25°C:

$$E = E' + 0.059 \log a_{Na^+} \; [V] \tag{31}$$

or introducing the pNa value $\equiv -\log a_{Na^+}$:

$$E = E' - 0.059 \; pNa \; [V]. \tag{32}$$

Since H^+ ions are always preferentially indicated by these types of glasses, the pH values of the test solutions should be about four powers of ten lower than the corresponding pNa or pK values. Figure 23 shows the logarithmic relationship between the

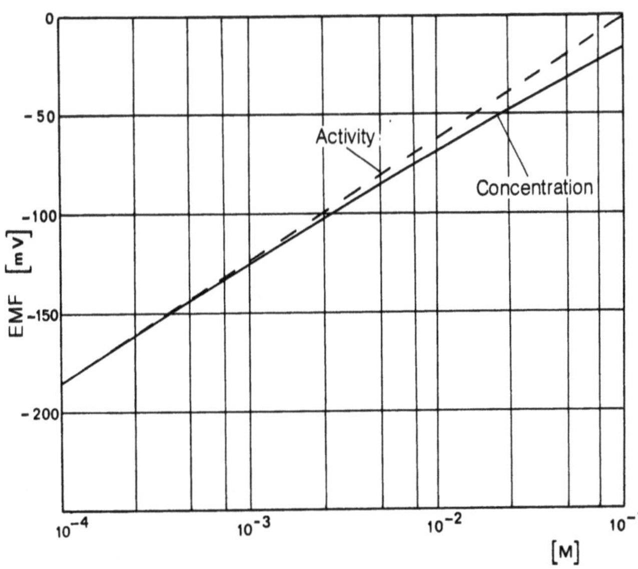

Fig. 23. Activity and concentration calibration curves for the sodium ion

EMF of a sodium-selective measuring cell and the corresponding sodium ion activity and concentration [77]. Figure 24 summarizes the different selectivities of a cation-sensitive glass [78].

3.2.1.3 Handling

Before a glass membrane electrode is ready for operation it requires a definite soaking time in water. The ion which is later to be measured should be added to this soaking solution (about 0.1 M). We have already seen in Chapter 1.3 why this is necessary for stable readings. For the reasons discussed there, ion-selective glass membrane electrodes are inclined to drift if suddenly transferred from a solution containing one cation to another containing a different cation. This effect is especially marked after a silver ion determination, because silver ions are only very slowly released from the gel layer. *If one wants to switch from one cation to another when using a cation-*

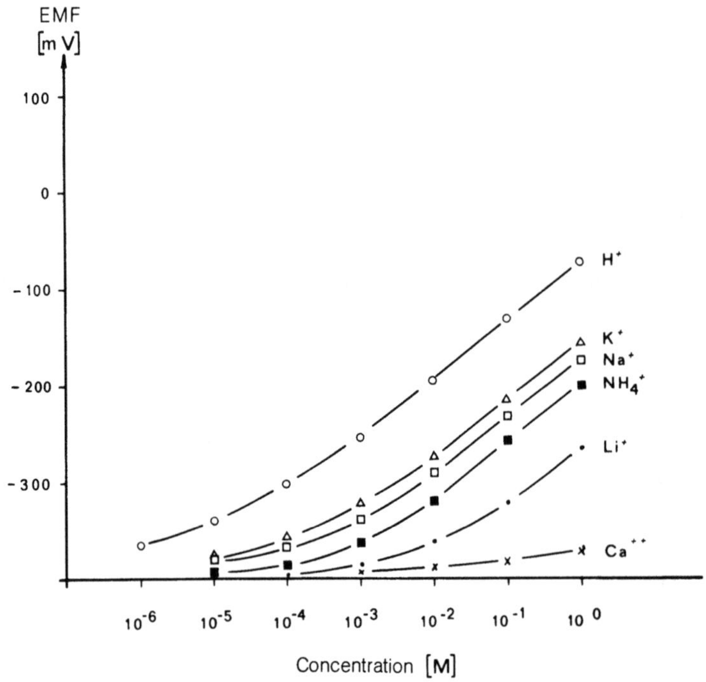

Fig. 24. Selectivity behavior of a cation-selective glass membrane electrode

sensitive electrode, a conditioning period of many hours is highly recommended to insure stable results. The ion-sensitive membrane should not be wiped off between potential measurements in different test solutions because this can easily damage the gel layer. A thorough rinsing with distilled water is quite sufficient. The diluting effect of the remaining water may be undesirable when working with small sample volumes. In such cases it is permissible to carefully blot the electrode dry with an absorbant paper towel. *Drying out of the electrode should be avoided.* When not in use, each glass electrode is best stored in a 0.1 M solution of its corresponding measured ion.

Even if the glass electrode is to be used in non-aqueous media, a sufficient soaking in water is still necessary. Following this the glass membrane electrode must be *conditioned with the appropriate sample medium.* It has been shown that ion-selective glass membrane electrodes retain their selectivity properties (excepting the H^+ ion) when transferred to partially aqueous systems (up to 90% ethanol, acetone, ethylene glycol or dimethylformamide). Any shift in potential relative to the pure water phase can be traced back to a change in liquid junction potential at the aqueous reference electrolyte/test solution phase boundary [79]. This potential shift is a serious problem for polarographic measurements in non-aqueous solvents. It can be overcome in our case, however, by introducing an ion-selective solid-state membrane electrode (electrodes with organic phases are seldom suitable), which can serve as a reference half-cell without liquid junction. During prolonged use in non-aqueous solvents, it is advisable to occasionally soak glass electrodes in water.

In contrast to reference electrodes, glass membrane electrodes require no special maintenance. Only in the case of combination electrodes does the salt bridge in the outer glass casing containing the reference element need to be refilled. Dirty glass membrane electrodes can be directly treated *for short time periods* with chromosulfuric acid. Albumin precipitates can be removed quite well with a pepsin-HCl mixture (5% pepsin in 0.1 M HCl). In both cases the electrodes must subsequently be thoroughly soaked in water. If the electrode function is completely and irreversibly impaired (a conclusion which should only be reached after testing the reference electrode), then it can occasionally be reactivated by a short etching with five percent hydrofluoric acid solution, buffing away the surface glass layers and soaking in water for many days.

The lifespan of glass membrane electrodes can reach many years. Most are mechanically destroyed much sooner through careless handling.

3.2.1.4 Sample Preparation

Since hydrogen ions are in large part potential-determining with all types of electrodes available to date, it is important to keep the pH value of the test solution about 4 units above analogous pNa or pK values. It is not necessary to precisely fix this pH value; only in the case of a cell without liquid junction using a pH glass membrane electrode or a hydrogen electrode as reference electrode should the pH value be fixed. A 1 M NH_4OH + 1 M NH_4Cl solution is suggested as a buffer solution [80]. For industrial applications NH_3 gas has proven useful. 2-amino-2-hydroxymethyl-1,3-propanediol has also proven successful for use in biological systems [81]. In this latter case, instabilities in the liquid junction potential have been noted when using reference electrodes with linen fiber diaphragms, so it is better to use a reference electrode with a double salt bridge (K^+ ions may interfere) and sleeve diaphragms. *Inorganic lyes, such as $Ca(OH)_2$, $Ba(OH)_2$, etc. are not to be recommended for adjusting the pH in trace analyses, because they are frequently contaminated with traces of monovalent cations.* Organic bases with large buffer capacities (such as diisopropylamine) are suitable, since they are much less often contaminated with inorganic ions. In spite of this, care must be taken if they are stored in glass bottles, because alkali ions can easily leach out of the glass walls. *Synthetic containers are much preferred for long-term storage of basic solutions.* The anion used does not influence the measurement; its only effect is to alter the activity coefficient of the corresponding cations. *Caution is advised for measurements in the presence of fluoride ions: if the solution is too acidic, the glass membrane may begin to dissolve.* For more exact procedures with different techniques, see Chap. 5.

Of all the types of glass membrane electrodes on the market, the sodium-selective electrode is the most widely used for direct sodium activity determinations. These measurements are as straightforward as pH measurements, since, except for the Ag^+ ion, all other species present in the test solution can be ignored. The cation-sensitive electrode is much less suitable for potassium ion determinations, although it has been used in the past only for lack of a better alternative. Friedman et al. [82] reports a simultaneous determination of both sodium and potassium in blood and plasma samples with the help of both types of electrodes. Here a computer was employed to automatically correct the measured potassium content based on the selectivity coefficients and the sodium level indicated by the sodium electrode. In his work an accuracy of within 0.5% is reported, better than that attainable with flame photometric analysis. With the same technique, however, Leonard et al. [83] achieved an accuracy of only ~ 10% for the determination of potassium in serum. Such discrepancies between measurements using the same types of electrodes can be explained by the use of different reference electrodes, and the resulting different electrolyte contacts. The cation-sensitive glass membrane electrode has lost importance since the introduction of the potassium-selective ion-solvating electrode. This electrode allows nearly interference-free, precise potassium determinations, even in the presence of varying amounts of sodium ions. The cation-sensitive electrode is now used predominately as an endpoint indicator in the titration of monovalent ions [78].

3.2.2 Homogeneous Solid-State Membrane Electrodes For Ag^+, Cd^{2+}, Cu^{2+}, Pb^{2+}, S^{2-}, F^-, Cl^-, Br^-, I^-, SCN^-, CN^- Ions

3.2.2.1 Principles

We have seen that a silver wire can be an electrode of the first kind, and will respond to silver ions in solution according to the Nernst equation, provided that the exchange current density for the phase transfer of Ag^+ ions is larger than that for any other ions in solution. For an interference of the first kind, an ion must be present which can gain more energy than a silver ion upon crossing the phase boundary and depositing on the silver support. This would be the case for all species which are nobler than silver. The electrode reaction:

$$Ag^+ + e^- \rightleftharpoons Ag^0 \qquad \text{(R11)}$$

is a redox-reaction, and as the equation shows electrons also take part, in contrast to the mode of operation of glass membrane electrodes. The potential of the silver electrode is also influenced by oxidizing or reducing agents since these can donate electrons to, or remove them from the electrode metal. Thus disturbances in the behavior of electrodes of the first kind can always be observed with redox agents; therefore they have not been able to be employed as selective sensors. A coating of sparingly soluble salt (electrode of the second kind) does not change this redox-sensitivity very much, since with the usual methods of preparation, it is impossible to eliminate all contact of the solution with the pure metal. In addition, electrons have the ability to "tunnel" through monomolecular layers[1].

[1] More recent investigations of C. Harzdorf [84] have shown that thick layers electrolytically deposited on silver (for example anodized for one hour at 1 mA/cm^2 in 1 M KCl, KBr, 0.1 M KI solution and finally clearly polished with a soft rag) are completely equivalent to the homogeneous solid-state membrane electrodes.

3.2.2.2 Construction

In 1937, Kolthoff and Sander [85] showed that a disk of fused silver chloride can be used as an ion-sensitive material in an electrode built in the same fashion as the pH glass membrane electrode. With an appropriate reference electrode this construction provides the same EMF as an AgCl coated silver wire, except it no longer exhibits a susceptibility to redox agents. Such electrodes do show definite light-sensitivities, and this stands in the way of their wider application. This light-sensitivity is understandable if the polycrystalline silver halide is viewed as a semiconductor [3]. In this case the band gap between valence and conduction bands is smaller than in a single crystal. Under certain circumstances it is small enough that light in the blue spectral region is sufficient to promote an electron from the valence band into the higher energy conduction band. The resulting "hole" in the valence band, as well as the electron in the conduction band, can either recombine or, under the proper energy circumstances, oxidize or reduce adsorbed materials. Fig. 25 schematically shows the oxidation of an

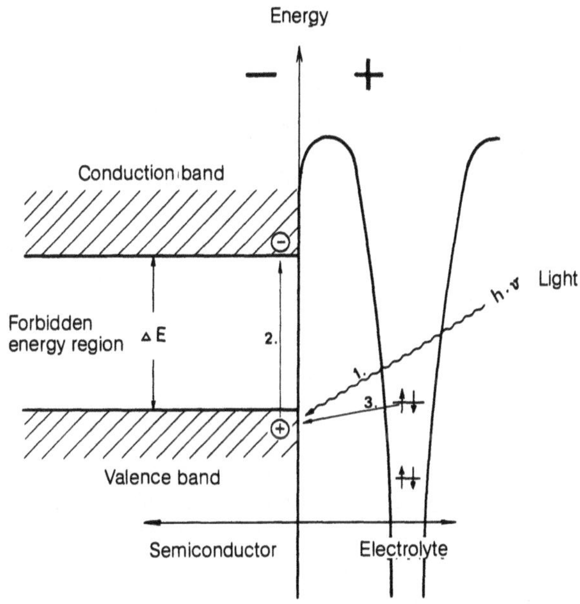

Fig. 25. Schematic explanation of the photo-EMF

adsorbed species through irradiation. The net effect of all the radiation-induced processes is reflected in a corresponding change in the electrode potential relative to the equilibrium Galvani potential in the dark. With electrodes containing a pressed membrane of Ag_2S this effect is not so pronounced. *Silver sulfide* is an ion-conducting material, as are many silver salts. At room temperature it exhibits a much larger silver ion conductivity than the corresponding halide compounds, and in contrast to the latter it can easily be pressed into thick disks. Because of its silver ion conductivity, this material shows a Nernstian response to silver ions in solution, and is thus well-suited for use in ion-selective electrodes. Such semiconductor-like materials do not respond so strongly to redox systems because of their larger forbidden energy bands (see Fig. 25) which cannot accommodate any electrons. The more energetically favorable valence band is full, so the only redox possibility is electron donation into the conduction band. Such a process requires a relatively high energy, so that only a few electrons in the solution redox system can be transferred. The net result is a general insensitivity to oxidizing or reducing agents.

Both single crystals and polycrystalline materials are equally suitable for electroactive phases. Chloride or bromide electrodes made from single crystals of AgCl or AgBr exhibit more favorable selectivity coefficients for Br^- and I^- than the polycrystalline pellets of Ag_2S mixed with the corresponding silver halides [86]. Here a different mechanism of establishing the potential must be assumed. In contrast to the mechanism sketched further below, it appears that the anion is also directly involved in the phase transfer with pure AgCl and AgBr membranes. The correspondingly higher exchange current densities are the source of the better selectivities.

A cupric selenide single crystal (as a good copper ion conductor) is the active phase in the Radiometer copper electrode; an Ag_2S single crystal (as a good silver ion conductor) is used by Crytur (Research Institute of Single Crystals, Turnov, CSSR) as a membrane in their silver and sulfide electrodes[1]. The solid-state membranes of Electrofact (Netherlands) consist of a mixture of gold powder ($< 3\ \mu m$)/silver salt 1:10 and are claimed to show no redox-sensitivity [88].

The fluoride electrode occupies a special place because of its outstanding selectivity. Here the active phase consists of a LaF_3 single crystal doped with Eu^{2+} to lower the ohmic resistance of the membrane. The unusual selectivity of this electrode rests on the fact that *only* fluoride ions are involved in the phase transfer [4].

Due to its large compressibility, silver sulfide serves, for many manufacturers, as a basis material for halide as well as metal electrodes. In this case only about 30% of the corresponding silver halide is used relative to silver sulfide, and the mixture is compressed at higher pressures and temperatures. It is, strictly speaking, no longer a homogeneous solid-state membrane.

As a rule, the commercially available solid-state membrane electrodes consists of a synthetic electrode body (PVC, epoxy, polypropylene, teflon, etc.), in which the electroactive single crystal or pellet is cemented (Fig. 20). Most contain an Ag/AgCl element dipping into an inner solution. In order to attain a constant Galvani potential at both inner phase boundaries (Ag/AgCl/inner solution/membrane), it is necessary to stock the filling solution with a constant amount of those ions involved in potential-creating processes at the interfaces. The chloride concentration is usually kept larger, so that the inner shunt electrode equilibrium Galvani potential corresponds more or less to that of the outer reference electrode (symmetric cell). With the fluoride electrode a solution about 1 M in NaCl and 10^{-2} M in NaF can be used. The manufacturers offer a filling solution with which a zero potential reading results when the electrode is dipped in a 10^{-3} M F^- solution and complemented with a calomel reference electrode.

[1] In this manufacturer's chloride electrode, the resistance of the AgCl single crystal used is lowered so far through doping with the O^{2-}ion, that problem-free measurements are possible [87].

3.2.2.3 Preparation of Solid-State Membrane Electrodes Based on Ag_2S

The starting material is Ag_2S powder of sufficient purity. Foreign ions can, under certain conditions, influence the properties of the semiconductor (for example the Fermi level). To eliminate these impurities as much as possible, it can be freshly precipitated from a 10^{-2} M $AgNO_3$ solution ($AgNO_3$ reagent grade; triply-distilled water) with the introduction of gaseous H_2S. The precipitate Ag_2S is washed with HNO_3 diluted with water 1:1 and plenty of hot water. After drying at 110°C the Ag_2S is treated with carbon disulfide to remove any elemental sulfur. It is then washed with ethanol and dried once again at 110°C. To produce a pellet about 3mm thick, the appropriate amount of Ag_2S is placed in a suitable pellet press (IR spectroscopy or X-ray fluorescence), compacted under vacuum for 10 minutes at 1,000 kg/cm^2, and then compressed for about 2 hours at 8,000 kg/cm^2. In preparing the corresponding metal electrodes, the 30% metal sulfide portion is best added to the nitrate before precipitation with H_2S. If brittle or cracked pellets result, the compression should be carried out at higher temperatures (> 150°C). The actual electrode is then constructed by cementing (Araldit, UHU plus or equivalent two component cement) the pellet in the end of plastic or glass tube of corresponding dimension. As an inner shunt system, a silver wire dipping in a 10^{-2} M $AgNO_3$ solution is used. In each case extreme care must be taken to insure that no direct contact exists between the inner solution and the outer sample solution (due to for example, poor cementing or tiny cracks in the membrane).

To improve the response time, the surface of the solid-state membrane electrode can be polished to a luster with diamond paste or moist Al_2O_3 of particle size $< 0.1\ \mu$. Poisoned electrodes can usually be restored by sanding and repolishing the surface.

3.2.2.4 Characteristics

Electrodes with pure Ag_2S membranes in connection with suitable reference electrodes respond to silver ions in solution according to the Nernst equation:

$$E = E' + \frac{R \cdot T}{F} \ln a_{Ag^+} . \qquad (33)$$

Because of the very low solubility product of silver sulfide

$$K_{S(Ag_2S)} = 6 \times 10^{-50} ,$$

a very large working range is to be expected. In practice, however, this range extends only from 1 M down to 10^{-7} M silver ion concentration. This is due to the large propensity for silver ions to adsorb on surfaces. *The detection limit of this electrode is restricted not by the intrinsic solubility, but rather by the instability of less than 10^{-7} M silver ion solutions.* The same problem arises here as was previously encountered with pH measurements of unbuffered solutions. If a stable solution in this concentration range is to be prepared, a buffer system must be found which can hold the silver ion activity at a certain level via an equilibrium reaction. The sparingly soluble silver salts make good silver ion buffers. If an excess of such a material is sitting at the bottom of a saturated solution, the silver ion activity in solution is determined by the solubility product of the precipitate. Irreversible adsorption can thus be compensated for with this system. Experiments show the response of this electrode strictly follows the Nernst equation in such solutions down to 10^{-20} M silver ion activities! (See Fig. 26) This activity corresponds to only two or three silver ions per ml of test solution!

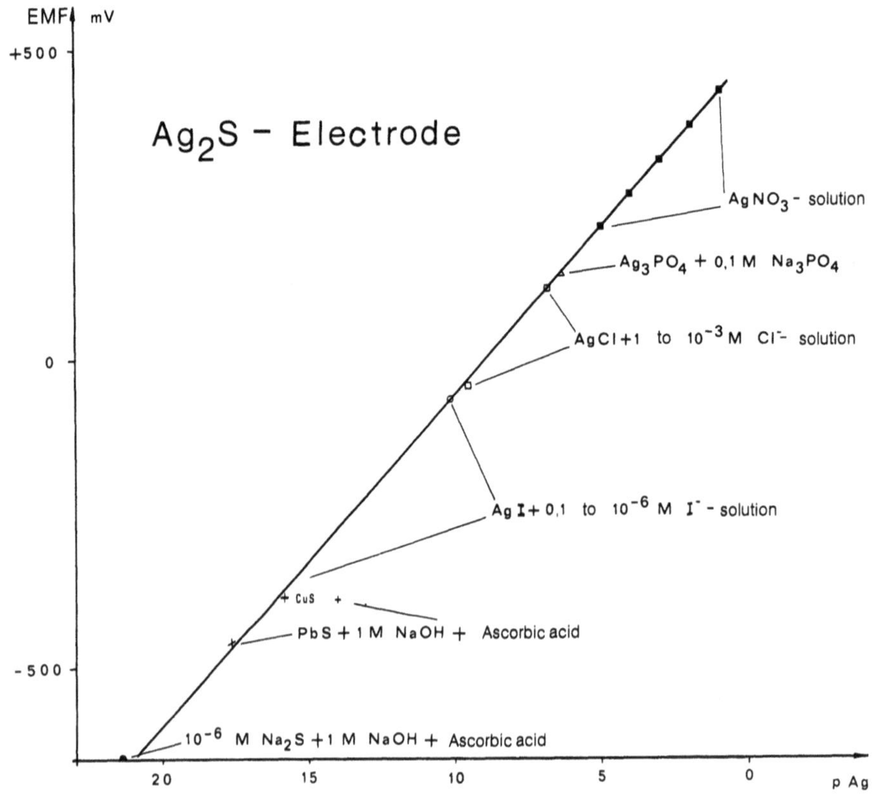

Fig. 26. Calibration curve for a silver-selective Ag_2S solid-state membrane electrode

The electrode material itself represents a silver ion buffer through the equilibrium:

$$2\,Ag^+ + S^{2-} \rightleftharpoons Ag_2S. \qquad (R12)$$

According to the law of mass action, the silver ion level is controlled by the sulfide ion activity. Thus it follows that the electrode also responds to free *sulfide ions.* As previously shown in the discussion of electrodes of the second kind, this potential dependence can be described by the equation

$$E = E' - \frac{R \cdot T}{2F} \cdot \ln a_{S^{2-}}\,. \qquad (34)$$

If the corresponding silver halides are used as silver buffers, then due to their larger solubility product the silver ion level will be higher than that due to Ag_2S alone. In this case the silver ion level will determine the potential, and according to Eq. (33) it will be more positive. However, according to the equilibrium:

$$Ag^+ + Hal^- \rightleftharpoons AgHal, \tag{R13}$$

the activity of the halide or pseudohalide ion (SCN^-) influences the free silver ion activity and, thus, the electrode potential:

$$E = E' - \frac{R \cdot T}{F} \cdot \ln a_{Hal^-} . \tag{35}$$

Electrode manufacturers add the appropriate silver halide to the Ag_2S support matrix in finely divided form and offer electrodes with these AgHal-Ag_2S mixed pellets as halide solid-state membrane electrodes.

In view of this mode of operation, it is easy to understand why those anions which form less soluble compounds with silver than does the measured ion interfere with these halide electrodes. The number of interfering ions decreases from chloride to bromide to iodide and finally to sulfide, since correspondingly fewer anions exist which can produce less soluble silver salts. Only Hg^{2+} interferes with the sulfide electrode. This is because Hg^{2+} forms an equally insoluble material with sulfide which can deposit on the electrode surface. In determining sulfide ion certain precautions are necessary. Sulfide is sometimes inclined to associate with hydrogen ions:

$$S^{2-} + H^+ \rightleftharpoons HS^-. \tag{R14}$$

In addition, traces of sulfide are easily oxidized by O_2 present in solution from contact with air.

Earlier, the silver ion activity in the sample solution was fixed by the addition of the appropriate *silver* halide. Since the electrode material also responds to sulfide ion, a sulfur buffer can also be utilized to attain a stable potential. Suitable buffer systems for sulfide are CuS, CdS and PbS which, due to their K_S values, establish a higher, stable S^{2-} ion level comparable to that of Ag_2S. Once again, according to the law of mass action, the buffered sulfide ion level is controlled by the corresponding metal ion. This means that changes in the Cu^{2+}, Cd^{2+} or Pb^{2+} activities affect the Ag^+ ion activity, and thus the electrode potential, via the equilibrium reactions:

$$M^{2+} + S^{2-} \rightleftharpoons MS \tag{R15}$$

$$2Ag^+ + S^{2-} \rightleftharpoons Ag_2S \tag{R16}$$

$$K_{S(MS)} = a_{M^{2+}} \cdot a_{S^{2-}} \tag{36}$$

$$K_{S(Ag_2S)} = a^2_{Ag^+} \cdot a_{S^{2-}} \tag{37}$$

$$a_{Ag^+} = \sqrt{\frac{K_{S(Ag_2S)} \cdot a_{M^{2+}}}{K_{S(MS)}}}. \tag{38}$$

In this way pellets which contain CuS, PbS or CdS in addition to Ag_2S behave as though they were Cu, Cd or Pb electrodes (electrodes of the third kind). By substituting Eq. (38) into Eq. (33) and combining the constant terms in E' the electrode potential can be expressed:

$$E = E' + \frac{R \cdot T}{2F} \ln a_{M^{2+}} . \tag{39}$$

In this case, all ions interfere which can form less soluble precipitates with the sulfide ion than the respective metal ion. Above all, it is the Ag^+ and Hg^{2+} ions which interfere.

A complex interference can result if Cu^{2+} and Cl^- ions are both present in *large concentrations.* Then the equilibrium between the electrode material and the Cu^{2+} and Cl^- ions in solution:

$$Ag_2S + Cu^{2+} + 2Cl^- \rightleftharpoons 2\,AgCl + CuS \tag{R17}$$

is shifted too far to the right by the law of mass action. This results in the build-up of an AgCl layer, which now responds to chloride ion rather than copper ion (because the silver ion buffer effect of AgCl produces a higher Ag^+ ion level than Ag_2S–CuS).

When using the Ag_2S–AgI membrane (iodide electrode) to detect cyanide in the range 0.1 to 10^{-8} M, another interference can arise through *silver complex formation* which results in dissolution of the membrane material. Possible electrode reactions are:

$$AgI + CN^- \rightleftharpoons AgCN + I^- \tag{R18}$$

$$AgI + 2\,CN^- \rightleftharpoons Ag(CN)_2^- + I^-. \tag{R19}$$

The equilibrium constant of this second reaction is so large ($\sim 10^4$) that only a vanishingly small CN^- ion concentration remains near the electrode surface. The $Ag(CN)_2^-$ formed diffuses into solution along with the iodide ion. Now, however, this free iodide determines the electrode potential, since once the electrode phase has started to dissolve the remaining phase responds to the Ag^+ ions in solution, whose activity is controlled by the iodide concentration. Two cyanide ions are required to liberate one iodide ion, so the Nernst equation for this reaction is:

$$E = E' - \frac{2R \cdot T}{F} \ln a_{CN^-} . \tag{40}$$

Instead of the usual 59.16mV (25°C) change for a 10-fold change in activity, if the predominant process is represented in reaction (R19) we would expect twice as large a potential. It must be pointed out, however, that this is an overly simplified picture. In reality, under steady-state conditions both species participate in the potential-determining process and are subject to the laws of diffusion. This leads to the following expression for the electrode potential [89]:

$$E = E' - \frac{R \cdot T}{F} \ln \left(\frac{D_{CN^-}}{D_{I^-}} a_{CN^-} + a_{I^-} \right). \tag{41}$$

Hence cyanide electrodes also exhibit slopes between 58 and 118mV/power of ten.

Because of the dissolution reaction (R19) the electrode should not be immersed in cyanide solutions above 10^{-3} M for long periods of time. It is clear that H^+ ions play an important role here, as they did for S^{2-} determinations. In too acidic solutions the volatile HCN forms and escapes from solution. This alters the cyanide concentration of the test solution, and moreover endangers the analyst!

A further interference can arise during the actual determination. The iodide electrode (AgI–Ag_2S) responds to *mercury ions* in the concentration range 10^{-4} to 10^{-8} M according to:

$$AgI + Hg^{2+} \rightleftharpoons HgI^+ + Ag^+. \tag{R20}$$

Here again the slope of the Nernst equation observed is not what one would expect for divalent ions. Since the Ag_2S electrode responds only to silver, and only one silver atom is released for each Hg^{2+}, a slope of 59.16mV (25°C) per tenfold activity change results [90]. In practice it is important to realize that reproducible values often can only be obtained if the electrode surface is sanded and polished after each measurement.

There have been many attempts to develop selective precipitate membrane electrodes for the important sulfate and phosphate anions. Unfortunately they have all failed to date. A sulfate-sensitive solid-state membrane electrode with a pellet of 31.7% Ag_2S, 31.7% PbS, 31.7% $PbSO_4$ and 5% Cu_2S has been described [91], but its selectivity over I^-, HPO_4^{2-} and SO_3^{2-} is low.

A pellet of 40 mol % HgS and 60 mol % Hg_2Cl_2 (compressed for 4 hours at 150°C, inner shunt-element of mercury) with a resistance of < 100 Ω for a few millimeter thick membrane has been described [9]. It is claimed to give a Nernstian response to chloride ions between 0.1 and 10^{-7} M with a selectivity over bromide and iodide ions $k^{pot}_{Cl-Br, I} \approx 3$, an appreciably better selectivity than the previously described Ag_2S–AgCl pellet [92].

Instead of using pellets mixed with the respective silver ion powder, these materials can simply be suspended in the solution to achieve the same selective behavior (see Fig. 26). A simple Ag_2S solid-state membrane electrode can then serve as the measuring electrode. Above all one must eliminate the presence of O_2 and be sure of the optimum pH value, since both of these factors affect the sulfide equilibria. Time constants can be very long with this type of set-up.

Because of the great importance of the fluoride electrode its properties will be discussed here in some detail.

The EMF of a cell containing fluoride ion solution, fluoride electrode and a suitable reference electrode follows the fluoride ion activity between 1 and 10^{-6} M according to the Nernst equation:

$$E = E' - \frac{R \cdot T}{F} \ln a_{F^-} \tag{42}$$

at 25°C this becomes:

$$E = E' - 0.059 \cdot \log a_{F^-} [V]. \tag{43}$$

The selectivity of the active phase (LaF_3 single crystal) is so great, that even a 1,000-fold excess of halide ions, NO_3^-, PO_4^{3-}, HCO_3^-, etc. is not detected; even cations do not interfere through direct indication. If the EMF changes with the addition of a foreign electrolyte to the sample solution, this can be traced to a corresponding change in the activity coefficients. This electrode owes its great popularity to its lack of interferences. The ability to indirectly determine a whole series of cations which form strong complexes with fluoride (such as Al^{3+}, Fe^{3+}, Ce^{4+}, Li^+, Th^{4+}, etc.) accounts for its widespread use. Even with thick single crystals, the resistance of the electrode is only about 0.15 to 5 MOhm.

An influence of the stirring rate (1 to 10mV) is noticeable when working with trace amounts of fluoride. This can be traced in part to the diffuse charge distribution at the electrode surface, and also in part to a corresponding change in the liquid junction potential at the reference electrode electrolyte/sample solution interface. This stirring rate dependence can be minimized by adding an inert electrolyte (for example 5 M $NaNO_3$) [93]. Measurements without stirring are inadvisable because of the longer response time. For exact work in the lower concentration range ($< 10^{-4}$ M) the same hydrodynamic conditions should be employed: same stirrer, stirring speed, container, solution volume and depth and angle of the measuring electrode.

The detection limit is governed by the solubility of the single crystal material in the test solution. Even if a solution contains no fluoride to start with, once the electrode is immersed a concentration of fluoride and lanthanum corresponding to the solubility product of LaF_3 will appear. Hence there will always be a small, constant amount of fluoride ion in solution, and since the electrode responds to this ion, the detection limit is determined[1]. In neutral solutions this corresponds to a working range of 1 to 10^{-6} M [93]. When working near this limit response times of a few minutes must be reckoned with. For this reason the length of time required to reach a stable electrode potential is longer when going from concentrated to dilute solutions than vice versa. Large excesses of species which are able to bind lanthanum ions such as citrate, PO_4^{3-}, HCO_3^-, etc. also slow down the response time[2]. The stability of the fluoride electrode response is excellent; it can vary as little as ±2mV per week [97].

---

[1] The detection limit can also be determined by irreversible adsorbtion on the electrode surface [94, 95].

[2] The time dependence of the potential following immersion of a fluoride electrode in solution lies in the msec to sec range through the following relation: $t/E_t = a + bt$ [96].

With measurements in cells without liquid junction a standard deviation of about 0.1mV can always be achieved. All these factors make this electrode particularly suited for constructing cells without liquid junction, since it can serve as a reversible reference electrode.

With fluoride-selective single crystal membrane electrodes only interferences of the second kind are observed which exert an influence on the solution equilibria of bound and free ions. It is known that H^+ ions combine with F^- ions to form undissociated species:

$$H^+ + 3\,F^- \rightleftharpoons HF + 2F^- \rightleftharpoons HF_2^- + F^- \rightleftharpoons HF_3^{2-}, \qquad (R21)$$

which are not indicated by the electrode. In this way the fluoride ion activity is reduced with decreasing pH values. In this case the electrode potential becomes increasingly more positive as shown in Fig. 27. On the other hand, a very high pH value is also not

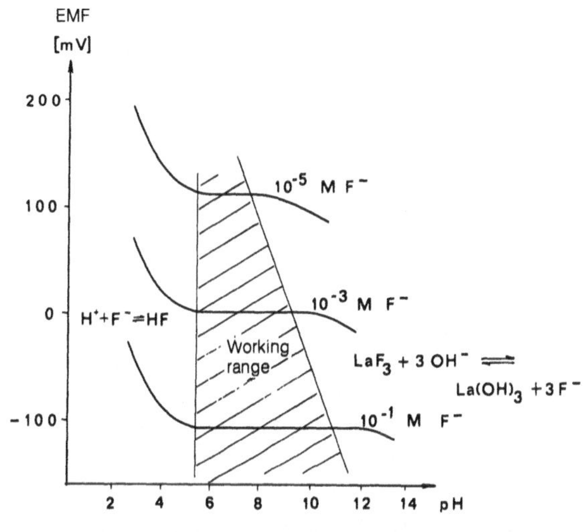

Fig. 27. pH dependence of the EMF between a fluoride-selective single crystal membrane electrode and a reference electrode

optimum. A layer of $La(OH)_3$ can build up on the electrode surface, since the solubility of $La(OH)_3$ corresponds roughly to that of the LaF_3 single crystal [58]. The reaction

$$LaF_3 + 3\,OH^- \rightleftharpoons La(OH)_3 + 3\,F^- \qquad (R22)$$

frees a distinct amount of fluoride ion. This is the reason why with solutions of low fluoride content the electrode potential becomes increasingly more negative with increasing pH [4]. This is also illustrated in Fig. 27.

Still unexplained is the reason why with some fluoride complexes, such as SiF_6^{2-} [98] or PO_3F^{2-} [99], the fluoride electrode responds as though completely or partially free fluoride were present.

The operational characteristics of homogeneous solid-state membrane electrodes are generally quite similar. As long as the inner shunt system is not an easily polarizable solid-solid contact [4], these electrodes can handle relatively high measuring circuit currents without deviating from Nernstian response (after correcting for the $i \cdot R$ potential drop).

All homogeneous solid-state membrane electrodes can be used as is in non-aqueous media as well. One must only take care that the electrode body is not attacked by the solvent. A large number of cells without liquid junction can be constructed in such media, and these are of great utility. The fields of organic electroanalysis and electrochemical synthesis are becoming increasingly more important thanks to the potentiostats now available. A number of the ion-selective electrodes discussed can also serve as reference electrodes for polarographic investigations in non-aqueous solvents.

Non-aqueous or partially aqueous systems are also preferred for precipitation titrations in order to lower the solubility of the resulting precipitate. Lower detection limits and more easily evaluated titration curves are thus obtained. The electrodes described here usually have electrical resistances less than 1 MOhm. This allows them to be used in *cooled* solutions (this resistance often has an exponential temperature dependence), which sometimes also lowers the solubility by a factor of 2 or 3 and thus extends the detection limit.

3.2.2.5 Handling

As a rule, the homogeneous solid-state membrane electrodes do not need to be conditioned. They can be stored dry and are immediately ready for use. Cleaning (sanding and polishing) is only necessary if the electrode has been poisoned by contacting a solution containing an ion which forms a less soluble precipitate with the measured ion than does the corresponding electrode counter ion.

The membrane surface should be kept free of scratches and deposits in the interest of a quick response time. Any scratches which eventually appear can be carefully removed with moist emery paper supplied by the manufacturer, and finally polished with diamond paste or Al_2O_3 ($<0.1\ \mu m$) on a cotton cloth.

To avoid collecting deposits, the tip of the electrode is coated with silicone oil using a cotton wad before the first use. This oil is also supplied by the manufacturer. *During all manipulations with the electrode surface extreme caution is required to insure that the seal of the membrane cemented into the electrode body is not damaged!*

The life span of solid-state membrane electrodes is determined by the frequency of repolishing. If the membrane surface does not need to be cleaned off very often, then the electrode can last for years. Electrodes with inner solutions should be filled enough to insure undisturbed contact between membrane and shunt element. If unseen air bubbles have formed inside the electrode between the membrane surface and the shunt electrode, irreproducible measurements may result, or under some circumstances the measuring circuit current may be interrupted. This is evident from unstable deflections of the ion meter instrument needle. Air bubbles can usually be released by mildly tapping the electrode while clamped in a vertical position (membrane below). In stubborn cases, especially with electrodes having small inner diameters, the electrode is quickly flicked downward, as a fever thermometer. If the electrode must be used in an inverted position with the membrane on top (as will be discussed later with micro-techniques), then it is especially important to be sure that the electrolyte contact between the inner membrane surface and the shunt element is not broken. In this case it is advisable to add agar-agar to the filling solution and completely fill the inner compartment with the gel.

Should the electrode manufacturers not give any indications as to filling solution composition, the following can be used: a 10^{-2} M $AgNO_3$ solution with all solid-state membrane electrodes based on Ag_2S; a 10^{-2} M $CuCl_2$ solution with copper ion-conducting membranes made of Cu_2S or cupric selenide; and a mixture of 0.1 M NaF and 1 M NaCl with the fluoride electrode.

3.2.2.6 Sample Preparation

Table 2 shows which ions interfere at what concentrations. A glance at Fig. 26 shows the same graphically. In preparing the sample it is important to take the proper pH range into account, not only because the hydrogen ions may take part in the electrode reaction directly, but also because they can combine with some ions to be measured to form undissociated species, such as HF, H_2S, HCN, HSCN, etc. The activities of the free anions can also be determined in acid media, of course, but they are very small due to the equilibria:

$$2\,H^+ + S^{2-} \rightleftharpoons H_2S \qquad (R23)$$

or

$$H^+ + CN^- \rightleftharpoons HCN, \text{etc.}, \qquad (R24)$$

and in these cases one does not determine the total content of the species of interest. Since these undissociated species are gases, they can easily escape from solution and alter the total anion content of the sample solution. At too high pH values, especially in the determination of copper, cadmium or lead, the corresponding metal hydroxide can precipitate out and lower the measured result. Thus the usable pH range is governed by the chemistry of the measured ion.

Special precautions are needed for the determination of the sulfide ion. At very low sulfide ion concentrations 1 M NaOH solution should be used. This insures that the sulfur present exists exclusively as the S^{2-} ion. In such solutions, however, the danger of oxidation by oxygen from the air is very great. The sample solution should be kept for just a short period of time, and in a well-stoppered flask without much air space above the solution. During the measurement the solution should not be stirred, and no inert gas should be introduced, since bubbling the solution will result in loss of gaseous H_2S. The electrode should be dipped as deeply as possible into the test solution, and allowed to stand undisturbed during the slow establishment of po-

Table 2. Homogeneous solid-state membrane electrodes

Electrode type	Active Phase	Potential-determining Ions	Working range [M]	Selectivity coefficient[1] $k^{pot}_{M\text{-}I}$	Recommended pH-range	Temperature range [°C]	[MΩ] at 25°C	Recommended reference electrode	Manufacturer
pAg	Ag_2S pellet	Ag^+, S^{2-},	1 to 10^{-7} $< 10^{-23}$, buffered	Traces of Hg^{2+} interfere, prolonged contact with Hg^{2+}-containing solutions requires surface treatment $Cu^{2+} \sim 10^{-6}$ $Pb^{2+} \sim 10^{-10}$	2 to 9	–5 to 100	< 1	Ag/AgCl with double salt bridge 1 M KNO_3	Beckman, Electrofact, Coleman, Corning, Foxboro, Metrohm, Orion, Philips, Polymetron, Radiometer, Schott & Gen., Tacussel
pAg	Ag_2S - single crystal	Ag^+, S^{2-}	s.above	$Cu^{2+} \sim 10^{-5}$; $Pb^{2+} \sim 10^{-6}$; $H^+ \sim 10^{-7}$; Hg^{2+} interferes	2 to 9			s. above	Crytur
pCu	CuS/Ag_2S	Ag^+, S^{2-}, Hg^{2+}, Cu^{2+}	1 to 10^{-8} $< 10^{-17}$, buffered	Ag^+, Hg^{2+} must be absent $Fe^{3+} \sim 10$; $Cu^{1+} \sim 1$; Cl^-, Br^- at higher concentrations	0 to 14	0 to 100	s. above	s. above	Beckman, Electrofact, Coleman, Corning, Metrohm, Orion, Tacussel, Philips
	Cu_2S	Cu^{2+}			3 to 14	0 to 70			Polymetron
	CuSe-single crystal	Cu^+, Cu^{2+}	1 to 10^{-6} $< 10^{-17}$ buffered	$Pb^{2+} \sim 10^{-4}$; $Cd^{2+} \sim 10^{-5}$; $Cu^+ \sim 10^{11}$; $Ag^+ \sim 10^6$; $Hg^{2+} \sim 10^4$	0 to 14	– 5 to 60	< 0.01	s. above	Radiometer
pCd	CdS/Ag_2S	Ag^+, S^{2-}, Cu^{2+}, Cd^{2+}	0.1 to 10^{-7} $< 10^{-10}$, buffered	Ag^+, Hg^{2+}, Cu^{2+} must be absent $Fe^{2+} \sim 200$; $Tl^+ \sim 120$; $Pb^{2+} \sim 6$ $Mn^{2+} \sim 3$	1 to 14	0 to 100	s. above	s. above	Electrofact, Metrohm, Orion, Philips
pPb	PbS/Ag_2S	Ag^+, S^{2-}, Cu^{2+}, Pb^{2+}	0.1 to 10^{-7} $< 10^{-10}$, buffered	Ag^+, Hg^{2+}, Cu^{2+} must be absent $Fe^{3+} \sim 1$; Cd ~ 1;	2 to 14	0 to 100	s. above	s. above	Corning, Electrofact, Metrohm, Orion

[1] according to the manufacturer

Table 2 (continued)

pS	Ag_2S– pellet	Ag^+, S^{2-}	1 to 10^{-6}, $< 10^{-17}$, buffered	Traces of Hg^{2+} interfere prolonged contact with Hg^{2+}-containing solutions requires surface treatment	13 to 14	–5 to 100	< 1	Ag/AgCl with double salt bridge 1 M KNO_3	Beckman Electrofact Coleman Corning Foxboro Metrohm	Orion Philips Polymetron Radiometer Schott & Gen. Tacussel
	Ag_2S– single crystal	Ag^+, S^{2-}	s. above	s. above Cu^{2+}~10^{-5}; Pb^{2+}~10^{-6}	s. above			s. above	Crytur	
pF	LaF_3– single crystal	F^-	1 to 10^{-6}	OH^-~0.1; other halides, NO_3^-, HCO_3^-, SO_4^{2-}~ $< 10^{-3}$	4 to 8	– 5 to 100	~0.2 to 5	usual reference electrode	Beckman Coleman Corning Crytur Metrohm Orion Philips	Foxboro Polymetron Radiometer
pCl	$AgCl/Ag_2S$– mixed pellet	Ag^+, Cl^-	1 to 10^{-5}	Br^-~10^2; I^-~10^6; OH^-~10^{-2}; CN^-~10^4; S^{2-} must be absent	2 to 11	~0 to 80	10 to 30	Ag/AgCl with double salt bridge 1 M KNO_3	Beckman Corning Electrofact Foxboro Metrohm	Orion Polymetron
	AgCl– single crystal	Ag^+, Cl^-	1 to 10^{-5}	Br^-~2; I^-~2; CN^-~ 8; OH^-~10^{-2}; NH_3~0.1	0 to 14	– 5 to 60	< 25	s. above	Radiometer Crytur	
	AgCl	Ag^+, Cl^-	1 to 10^{-5}	Br^-~1; I^-~10^2; CN^-~400; $S_2O_3^{2-}$~60 OH^-~10^{-2}; CO_3^{2-}~10^{-3}	1 to 10	0 to 100	< 1	s. above	Philips Schott & Gen. Tacussel	
pBr	$AgBr/Ag_2S$– mixed pellet	Ag^+, Br^-	1 to $5 \cdot 10^{-6}$	I^-~$5 \cdot 10^3$; CN^-~10^2; Cl^-~$5 \cdot 10^{-3}$; OH^-~10^{-5} S^{2-} must be absent	2 to 12	s. above	s. above	s. above	Beckman Corning Electrofact	Foxboro Metrohm Orion

Table 2 (continued)

Electrode type	Active Phase	Potential-determining Ions	Working range [M]	Selectivity coefficient[1] $k^{pot}_{M\text{-}I}$	Recommended pH-range	Temperature range [°C]	[MΩ] at 25°C	Recommended reference electrode	Manufacturer
	AgBr–single crystal	Ag^+, Br^-	1 to 10^{-6}	I^-~2; CN^-~1; Cl^-~$5\cdot10^{-3}$; OH^-~10^{-4}, NH_3 ~ $4\cdot10^{-3}$;	0 to 14	–5 to 60	< 25	s. above	Radiometer
	AgBr	Ag^+, Br^-	1 to 10^{-6}	I^-~20; CN^-~25; Cl^-~$6\cdot10^{-3}$; OH^-~10^{-3}; $S_2O_3^{2-}$~1.5; CO_3^{2-}~$2\cdot10^{-3}$	1 to 11	0 to 50	$\lesssim$ 1	s. above	Philips, Schott & Gen., Tacussel
pI	AgI/Ag_2S–mixed pellet	Ag^+, S^{2-}, I^-	1 to $5\cdot10^{-8}$	S^{2-}~30; $S_2O_3^{2-}$~$3\cdot10^{-2}$; CN^-~10^{-2}; Br^-~10^{-4} Cl^-~10^{-6}; SCN^-~10^{-4} OH^-~10^{-7}; NH_3~$3\cdot10^{-5}$	0 – 14	0 – 80	< 0.5	s. above	Beckman, Electrofact, Corning, Foxboro, Metrohm, Orion, Radiometer
	AgI	Ag^+, I^-		CN^-~0.34; CrO_4^{2-}~$4\cdot10^{-3}$ $S_2O_3^{2-}$~$7\cdot10^{-4}$; CO_3^{2-}~10^{-4} Br^-~$6\cdot10^{-5}$; Cl^-~$6\cdot10^{-6}$ S^{2-} must be absent	1 – 12	0 – 50	< 0.5	s. above	Crytur, Philips, Schott & gen., Tacussel
pSCN	AgSCN/Ag_2S–mixed pellet	Ag^+, SCN^-	1 to $5\cdot10^{-6}$	I^-~10^3; Br^-~10^2; CN^-~10^2; $S_2O_3^{2-}$~10^2; NH_3~10; Cl^-~0.1 OH^-~10^{-2} S^{2-} must be absent	2 to 12	0 to 95	< 100	s. above	Metrohm, Orion
pCN identical with pI	AgI/Ag_2S–mixed pellet	Ag^+, I^-, CN^-		I^-~10^2; Br^-~10^{-4} Cl^-~10^{-6}; OH^-~10^{-8} S^{2-} must be absent	11 to 13	0 – 80	< 30	s. above	Beckman, Electrofact, Corning, HNU, Metrohm, Orion, Polymetron, Radiometer
	AgI	Ag^+, I^-, CN^-		I^-~3; CrO_4^{2-}~10^{-2}; $S_2O_3^{2-}$~10^{-3}; CO_3^{2-}~10^{-4}; Br^-~10^{-4}; Cl^-~10^{-5}	10 – 12	0 – 50	0.5	s. above	Crytur, Philips, Schott & Gen.

[1] according to the manufacturer

tential. For trace analysis of sulfur the addition of a small amount of ascorbic acid or formaldehyde to reduce the amount of dissolved oxygen is recommended [80].

As Fig. 27 shows, the working pH range for the fluoride electrode lies between 5.5 and 6.5. This is easily attained with an acetate or citrate buffer [80]. At this pH value the *free* fluoride ion activity can be determined between 1 and $< 10^{-6}$ M.

If one is interested in the *total fluoride content* of a sample (including that portion of fluoride which is complexed with cations such as Al^{3+}, Fe^{3+}, etc.), then a complexing agent C must be added which can complex the cations of concern even more strongly than fluoride. Then the exchange reaction, for example:

$$AlF_6^{3-} + C^{3-} \rightleftharpoons Al{-}C + 6\,F^-, \qquad \text{(R25)}$$

will set the fluoride free so that it can be indicated by the electrode. Citrate, ethylenediaminotetraacetic acid (EDTA), ethylene glycol bis-(aminoethylenether)-N,N,N′,N′-tetraacetic acid and 1,2-diaminocyclohexane-N,N,N′,N′-tetraacetic acid are all suitable complexing agents. To minimize the work involved in a total fluoride content determination, the sample is diluted only once with a single solution containing the buffer, a salt to fix the ionic strength (NaCl, KNO_3, etc.) and the complexing agent. Such conditioning solutions can be obtained from the electrode manufacturers.

The solution which Orion offers can be made up (according to their own instructions) as follows [100]:

57 ml acetic acid, 58 g NaCl and 4 g of the disodium salt of 1,2-diaminocyclohexane-N,N,N′,N′-tetraacetic acid are added to 500 ml distilled water in a 1 l beaker. [The disodium salt is obtained from 36 g of the acid (Titriplex IV, Merck; Idranal IV, de Haen) in 200 ml 1 M NaOH; 18 ml of this solution corresponds to about 4 g of the salt.] The pH of the resulting solution is adjusted to between 5 and 5.5 with NaOH or HCl. It is then diluted to 1 l with distilled water. Naturally, all the chemicals used must be free of fluoride ion.

With such conditioning solutions sample preparation is simple: The test solution is diluted with the conditioning solution from 1:1 to 1:10, and all is ready for the measurement. Interferences seldom appear; only in the cases of extremely high concentrations of complex-forming cations (for example fluorspar analysis) or extremely acidic or basic solutions. In such cases additional complexing agent must be added, and the pH corrected afterward with a stronger acid or base. When working near the detection limit, complexing agents can sometimes cause problems because they also react with the LaF_3 single crystal, thus setting fluoride ions free according to reaction (R22). With large excesses of complexing agent this can raise the lower detection limit and slow down the response time of the electrode. *If very small traces of fluoride are to be determined, the excess of complexing agent should be kept to a bare minimum.*

In general, trace analyses are difficult because the extremely dilute solutions are only stable for short time periods due to irreversible adsorbtion on the container walls. Silver and fluoride solutions are among the worst; even 10^{-5} M solutions are so unstable in glass vessels that they last only a few hours before the concentration decreases. It is better to use plastic containers which are "conditioned" with the test solution for about 10 minutes. Just before the actual measurement, this is refilled with fresh sample solution. When progressing from concentrated to very dilute solutions, even adsorbtion on the electrode membrane itself can cause readings to drift and yield erroneous results. This requires repolishing the electrode surface, in some cases (such as Hg^{2+} measured with an iodide electrode) after every measurement!

In all cases a reference electrode with a double salt bridge is recommended, since many indicated ions form sparingly soluble salts (Ag_2S, HgS, CuCl, etc.) with the ions in the usual reference electrode filling solutions (K^+, Cl^-, Ag^+, Tl^+ or Hg^+). These salts can precipitate in the diaphragm and lead to unstable liquid junction potentials, if not complete breakdown of the reference electrode (see Chap. 2.3.3). If strongly alkaline solutions must be used, it is advantageous to adjust the pH of the outer salt bridge electrolyte to the same value as that of the sample solution in order to keep the liquid junction potential small (see Eq. 27). The same applies when working with strongly acidic solutions.

3.2.3 Heterogeneous Solid-State Membrane Electrodes For Ag^+, Cl^-, Br^-, I^-, CN^-, SCN^-, S^{2-} Ions

3.2.3.1 Construction

Even before the development of homogeneous precipitate membrane electrodes attempts were made to separate the functional cover layer of electrodes of the second kind from the supporting material in order to avoid the interferences from oxidizing or reducing agents. Not all precipitate compounds can be pressed or fused to mechanically stable materials, however, so an alternative method had to be developed to produce mechanically stable, electrolytically thick disks of these compounds.

The *heterogeneous* precipitate membrane electrodes stem from the work of Habers [101] and Marshalls [102]. Habers tried to attain this mechanical stability by imbedding the polycrystalline or amorphous compounds in paraffin wax. To increase the mechanical strength Tendeloo et al. [103] included cotton fibers. Since the active material is now heterogeneous, i.e. mixed in with an inert matrix, these are referred to as heterogeneous membrane electrodes. A few electrodes made in this way show some ion-sensitive behavior toward the ion of the sparingly soluble salt in the active phase, but the selectivities are usually poor. This is in part due to the cotton fibers, which can act as salt bridges between inner and outer solutions. Other support materials were tested, such as PVC, polyethylene, etc., but the inelasticity of the resulting membranes often rendered them prone to cracks and flaws.

Pungor et al. [104–109] found a suitable carrier material in *silicone rubber.* Silicone rubber is hard, yet elastic. It is water-repelling, but in spite of this a thin surface layer is capable of swelling. Pungor developed the technique of producing these membranes to the manufacturing stage, so the heterogeneous precipitate membrane electrodes currently available from Rakelkis, Budapest are based on his work. Amel (Milan) also produces heterogeneous solid-state membrane electrodes [110]. This firm has imbedded the appropriate silver halide in a polymer matrix (polyethylene). These commercially available heterogeneous solid-state membrane electrodes (Table 3) are built with an inner solution just as the corresponding homogeneous solid-state membrane types.

In more recent times the widest variety of sparingly soluble materials have been tested as active phases. A few whose ion-selective electrode functions have been investigated are: K_2SiF_6, $Ag_4[Fe(CN)_6]$ [111], Ag-halides [105], Ag_2S [109], CaF_2 [112], Ca-oxalate [113], Ca-stearate [114], $BaSO_4$ [41, 42, 115, 116], $BaCrO_4$ [116], K-tetraphenylborate [117], $PbWO_4$ [111], Mn(III)PO_4 [105], Mn(II)-oxalate [118], $MnCO_3$, $MnHPO_4$ [42], Al-8-oxyquinoline [105], Ni-diacetyldioxime [42, 114, 118], $Ni_3(PO_4)_2$ [42], Ni(II)-oxalate [118], $Co_3(PO_4)_2$ [42, 118], LaF_3 ThF_4 [112].

Usually the results were disappointing. Often all cations or anions of the same valence were indicated with no differentiation. The only materials to show ion-selective behavior were the silver halides and silver sulfide for Ag^+ ions, and CaF_2 and LaF_3 for the F^- ion. It is yet to be explained why the heterogeneous membranes only exhibit ion-selective behavior if they do not contain too much supporting material. In a more recent study [42] the authors, whose goal was to find new electrode materials for the transition elements, came to the conclusion that the crystalline form, the water content of the crystal and the counter ion all exert no great influence on the electrochemical behavior of the electrode toward the measured ion. All evidence indicates that only ionic salts, which have a good conductivity and allow a large exchange current density with ions in solution, exhibit the desired effects. From this it can be concluded, as with glass electrodes, that the actual potential at the gel layer/solution phase boundary arises because of the different chemical potentials in these two phases. Ion-exchange membranes can also be used as active phases [119, 120], but here too the selectivity is very poor. Electrodes of this type can only be used in pure solutions of a particular ion.

3.2.3.2 Preparation

The following procedure has been described in the literature [42]: The electroactive material is thoroughly mixed with a sufficient amount (50 weight %) of silicone rubber cement (for example "Clear Seal", a rubber vulcanized at room temperature, by General Electric, U.S.A.) and pressed to an approximately 0.5 mm thick membrane between two flat plates. After hardening in air, a piece is cut out with a cork-borer and fixed into the end of a suitable borosilicate glass tube with some additional silicone rubber cement. This step requires much care, since any contact between the inner solution and the sample solution will destroy the ion-selectivity.

According to another procedure, 70 wt % of the active phase (Ag halide) was mixed with 30 wt % dimethylpolysiloxane (Silopren K 1 000, Farbenfabrik Bayer) and polymerized at room temperature [121].

3.2.3.3 Characteristics

Since they operate in the same fashion, the electric and electrochemical properties of the heterogeneous solid-state membrane electrodes on the market today resemble those of the homogeneous precipitate membrane electrodes in many ways. The most important data are summarized in Table 3. By the same principles outlined in the previous chapter, the heterogeneous iodide electrode can also detect cyanide. In contrast to the homogeneous membranes, those which include silicone rubber as a support ma-

Table 3. Heterogeneous solid-state membrane electrodes

Electrode type	Active Phase	Potential-determining Ions	Working range [M]	Selectivity coefficient[1] $k_{M\text{-}I}^{pot}$	Recommended pH-range	Temperature range [°C]	[MΩ] at 25°C	Recommended reference electrode	Manufacturer
pAg	Ag_2S	Ag^+, S^{2-}	1 to 10^{-7} 1 to 10^{-17} buffered	Hg^{2+} interferes in traces	2 to 8	0 to 50	~1	Ag/AgCl with double salt bridge	Amel Radelkis
pCd	CdS	Ag^+, Hg^{2+}, Cd^{2+}	10^{-1} to 10^{-6}	Pb^{2+}~1; Zn^{2+}~10^{-3}; Co^{2+}~10^{-3}; Ni^{2+}~10^{-3}	1 to 8	0 – 50	s. above		s. above
pCu	CuS	Ag^+, Hg^{2+}, Cu^{2+}	1 to 10^{-6}	Pb^{2+}~10^{-3}; Cd^{2+}~10^{-3}; Co^{2+}, Ni^{2+}, Zn^{2+}~10^{-3}	1 to 14	0 – 50	–		s. above
pPb	PbS	Ag^+, Hg^{2+}, Pb^{2+}	10^{-1} to 10^{-6}	Cd^{2+}~1; Fe^{2+}~10^2	2 to 7	0 – 50	–		s. above
pS	Ag_2S	s. pAg	s. pAg	s. pAg	13 to 14	s. above	s. above		s. above
pCl	AgCl	Ag^+, S^{2-}, I^-, Br^-, Cl^-	1 to $5 \cdot 10^{-5}$	CN^-~400; I^-~86.5; $S_2O_3^{2-}$~60; Br^-~1.2; NO_3^-~1; OH^-~$2 \cdot 10^{-2}$; SO_4^{2-}~10^{-2}	2 to 12	s. above	s. above	1 M KNO_3	s. above
pBr	AgBr	Ag^+, S^{2-}, I^-, Br^-	1 to 10^{-6}	CN^-~25; I^-~20 $S_2O_3^{2-}$~1.5; NO_3^-~1; Cl^-~$6 \cdot 10^{-3}$	2 to 12	s. above	s. above		s. above
pI	AgI	Ag^+, S^{2-}, I^-, CN^-	1 to $5 \cdot 10^{-8}$	NO_3^-~1; SO_4^{2-}~1; CN^-~0.34;	3 to 12	s. above	s. above		s. above
pCN	AgI	s. above	10^{-2} to 10^{-6}	I^-~3; NO_3^-~3; SO_4^{2-}~3	11 to 13	s. above	s. above		s. above
pSCN	AgSCN	Ag^+, S^{2-}, I^-, Br^-	1 to 10^{-5}	I^-~70; Br^-~1;	2 to 12	s. above	s. above		s. above

[1] according to the manufacturer

terial form an aqueous gel layer (its formation can clearly be followed by the decrease in resistance of the membrane as the gel layer appears). This means that electrodes of the silicone rubber type should be conditioned before use. The somewhat longer response times of these electrodes relative to their homogeneous analogues [122] are also connected with this gel layer. Malissa and Jellinek [123] reported that a bromide electrode which had been poisoned with iodine regenerated so completely in only 10 minutes, that an interference-free determination of bromide was possible.

Caution is required when working with non-aqueous solvents to insure that the solvent does not dissolve the carrier material of the active phase (silicone rubber, polyethylene, etc.).

3.2.3.4 Handling

The commercially available heterogeneous solid-state membrane electrodes should be conditioned in a ~ 0.1 M solution of the ion to be measured for about 2 hours before the first measurement. With self-made electrodes having materials which are not ion-conductors, a conditioning in a 1 M measured ion solution at ~ 90°C is recommended. Drying out of the active phase is to be avoided, since the accompanying shrinkage can cause fine cracks in the membrane. Such electrodes are best handled like glass membrane electrodes, and stored in a 0.01 to 0.1 M solution of the indicated ion between measurements.

Caution is recommended when working with suspensions, so that sharp particles do not damage the relatively soft surface of the membrane. Poisoned electrodes can be regenerated with long periods of conditioning. In all other respects, including sample preparation, these electrodes closely resemble the homogeneous type.

3.3 Porous and Nonporous Supported Ion Exchanger and Neutral Carrier Membrane Electrodes

3.3.1 Ion-Exchangers For Ca^{2+}, Me^{2+} Cations and Cl^-, ClO_4^-, NO_3^-, BF_4^- Anions

In the electrode types discussed so far, a glass or crystalline matrix served to energetically stabilize the ions transferring from the solution phase into the electrode phase. These materials (glass, crystals or microcrystalline precipitates [3]) can be thought of as specific ion-exchangers with fixed counterions. Many attempts to produce selective electrodes based on other inorganic or organic solid ion-exchangers have failed. Sollner et al. [124, 125] found a suitable, easily workable electrode material for monovalent cations in collodion. For fixed counterions, carboxylate groups could be produced in very small amounts ($< 10^{-6}$ mol/cm^2) on the pre-cast collodion membrane (the end of a test tube covered with a collodion-solvent mixture and dried in a specified humidity) by a short oxidation in alkaline solution. If the collodion was mixed with Protamin, the resulting membrane responded to monovalent anions. In neither case was the response selective.

In liquid organic phases with corresponding ion-exchange properties the problem of differing ionic mobilities in the electrode material (Chapter 1.4) is solved. Once an ion has penetrated into the fluid organic phase, it can move freely regardless of how strongly it interacts with a counterion. In this case the selectivity must be achieved through a favorable position of the corresponding phase equilibrium:

$$\begin{aligned} &\text{Interfering Ion}_{(\text{solution})} + \text{Measured Ion}_{(\text{organic phase})} \\ &\quad \rightleftharpoons \text{Interfering Ion}_{(\text{organic phase})} \\ &\quad + \text{Measured Ion}_{(\text{solution})}. \end{aligned} \tag{R26}$$

For ion-specific behavior this equilibrium should lie completely to the left. Interfering ions should be excluded from entering the organic phase. As early as 1906 Cremer [8] described electrodes containing phenol and nitrobenzene saturated with picric acid as liquid ion-exchange phase. These respond to monovalent cations, but the selectivity is very slight.

At this point let us once again formulate the requirements for good ion-selective behavior based on the considerations in Chapters 1.3 and 1.4:

a) First of all, in order for a jump in chemical potential, a *single* species must be present in two sufficiently different phases. This requirement can be fulfilled if a water-insoluble liquid is chosen for the second phase. In order to avoid evaporation losses and corresponding changes in the electrode properties, the vapor pressure of the second phase should be as small as possible. Long-chain alcohols and aromatic compounds can satisfy this criterion.

b) This organic phase must exhibit a definite chemical potential of the indicated ion, so that the uptake or release of ions upon immersing the electrode in solution does not alter the identity of the electrode phase (ion buffer). Such a change would result in a drifting of the measured potential. In order to bring reasonable amounts of the indicated ion into the active phase before the measurement, oppositely charged groups must be present in this organic phase (requirement of overall electroneutrality). These counterions are also subject to some restrictions. They must form only sparingly water-soluble salts with measured ion, or they will be dissolved out of the organic phase upon contact with water. This would result in an unrealistically high concentration of the measured ion in the test solution, corresponding to a poor limit of detection for the electrode. Thus the measured ion-counterion association complex must be sufficiently soluble in the organic phase in spite of its low dielectric constant. This can be accomplished with counterions having long-chain pendant groups. For a specific response to a single ion, one is searching for an organic compound with lipophilic side groups, and a functionality which is known to have the ability to bind quickly and selectively to the measured ion.

In 1967, Ross [126] described an electrode based on a liquid ion-exchanger consisting of the calcium salt of dodecylphosphoric acid dissolved in dioctylphenylphosphonate. This electrode selectively responds to the calcium ion activity of the solution in the concentration range 0.1 to 10^{-4} M. This phosphorus compound was chosen because previous analytical chemistry had shown that phosphate and polyphosphate anions form stable complexes with Ca^{2+} ions. A diester was chosen to avoid the formation of mixed complexes between Ca^{2+} and H^+, which would result in an interference from the latter. The selectivity relative to Zn^{2+}, Fe^{2+} and Pb^{2+} is not very great, but these ions are not usually present in large enough quantities to interfere for most uses in physiology.

If the calcium complex is dissolved in a less polar medium, such as decanol, then all divalent cations are indicated with virtually no selectivity [127]. This exchange system is used as the active phase in electrodes used to determine water hardness.

Ross and Frant [128] describe an exchanger phase based on a substituted thioglycolic acid $R - S - CH_2 - COO^-$, which reacts selectively with Cu^{2+} and Pb^{2+} ions. Since the introduction of solid-state electrodes for these two ions, the development of liquid ion-exchanger systems for each individual ion has not been pursued.

If the counterion-containing molecule is positively charged, then an anion-selective behavior results. Thus exchanger systems containing Fe(phenanthroline)$_3^{2+}$ respond to ClO_4^- and NO_3^- ions. Tris(4,7-diphenyl-1,10-phenanthroline)Ni^{2+}-nitrate in p-nitrocymene system can be used to make a nitrate-sensitive electrode. If the NO_3^- anion is replaced with the BF_4^- anion in this system (this can be accomplished by shaking the ion-exchanger in the NO_3^- form with a $NaBF_4$ solution), then an electrode containing this exchanger responds to BF_4^- [129]. A response to the corresponding positive charge carriers Ni^{2+} and Fe^{2+} could not be observed. As already mentioned in Chapter 1.3, due to the charge delocalization in the metal-phenanthroline complex, the activation energy is too high to allow the rapid establishment of an equilibrium of the metal ion between the organic and aqueous phases. Chloride-selective behavior can be obtained by introducing a long-chain tetraalkylammonium ion (such as dimethyldistearyl ammonium chloride) as carrier of the counter-charge [127].

3.3.2 Neutral Carrier Compounds For Li^+, Na^+, K^+, Rb^+, Cs^+, NH_4^+, NR_4^+, Ca^{2+} and Ba^{2+} Cations

According to the ion-exchange principle, the transfer of a solvated ion from the solution phase into the organic phase is energetically possible only if the transfer is coupled with an association of the measured ion with the special counterion already present in the organic phase. This can be viewed as a kind of specific extraction by the organic exchanger.

In 1964, Moore and Pressman [130] discoveıed that certain neutral macromolecular compounds, such as the antibiotics, exhibit both cation-solvating and lipophilic behavior. Simon and Stefanac [43] described the first liquid membrane electrode with an antibiotic as the ion-selective phase. Simon et al. [131, 132] investigated the selectivity characteristics of nonactin and valinomycin towards alkali metal ions. These investigations led to the construction of a potassium-selective electrode. The antibiotics were dissolved in a nujol/2-octanol mixture or in diphenyl ether. The structure of these antibiotics (nonactin, trinactin, valinomycin, nigericin, etc.) provides for the formation of complexes in which the cations are contained in molecular cavities. These cavities are stablized by ion-dipole forces (s. Fig. 12). The extraordinary specificity of biological cells for certain ions is presumably based on this property. If these molecules are arranged in a very thin layer (biological dimensions, < 60Å), the charging of the organic phase associated with the cation-solvating behavior of these molecules cannot completely inhibit the transfer of these special cations through this layer. In order to construct stable electrodes thicker layers are required, so the charging of this phase now effectively inhibits further diffusion of the special cation. The excellent selectivity of the antibiotics in differentiating Na^+ and K^+ ions led to the investigation of synthetic molecules of similar structure. Pederson [44] found a similar selective behavior with macrotetralides[1] (Fig. 13). Simon [134] enumerates the problems encountered in trying to utilize these extremely specific cation-solvating, "tailor-made" molecules in practical electrode constructions. Often the organic phase solubilizes many ions and ion-pairs, so that the selectivity coefficients are many orders of magnitude less favorable than those theoretically possible (based on the experimentally obtained complex formation constants). The organic phase must be able to dissolve a certain amount of the measured ion along with its anion during the conditioning period in order to establish a definite chemical potential for this ion in the active phase (retention of phase identity, see Chap. 1.3). Kedem [135] was able to show that using millipore filter membranes as the supporting material, the active phase contained 10^{-4} M negatively-charged species[2].

3.3.3 Construction

Ion-Selective electrodes with liquid electro-active phases require special construction techniques. The rather unstable interface between two immiscible liquids must be stabilized, so that hydrostatic and hydrodynamic fluctuations in the sample solution do not exert too large an influence on the spacial charge distribution near the interface. There are a number of ways in which this can be accomplished.

In the first such electrode manufactured by Orion, the end of a glass tube was tightly closed with a cellulose dialysis membrane. Such membranes exhibit equally good permeabilities for most small ions. The glass tube contained a layer of liquid ion-exchanger up to 5 mm deep. An Ag/AgCl shunt element can be used in an agar-agar stabilized solution 0.01 M in the measured ion and in chloride ion (see Fig. 28). Such an electrode construction is easily self-made, although it is difficult to handle in practice. Its sensitivity to mechanical shaking is very large. In addition, the poor exchanger conductivity leads to long response times and slow measurements. With thinner exchanger layers, there is always the danger that the reference element solution will come into direct contact with the dialysis membrane, resulting in undifferentiated ion transport between this solution and the test solution and complete loss of the ion-selective behavior. Direct contact with the test solution can be hindered by making the tip of the electrode hydrophobic. Suitable materials for this are, for example, Desicote (Beckman) or dichlorodimethylsilane in carbon tetrachloride.

The series 92 electrodes of Orion represent a mechanically and hydrodynamically more stable electrode construction. In this series the organic phase is stabilized by a slice of hydrophobic membrane, as shown in Fig. 28. The exchanger liquid is held quite firmly in the pores of this membrane. Loss of the exchanger liquid through evaporation or dissolution in the sample solution is compensated for by replenishment

---

1 In addition to the adaptation of the molecular cavity to the radius of the trapped ion (which determines the thermodynamic equilibrium), the flexibility of the macromolecular framework also has an influence on the selectivity [133]. This once again indicates the importance of the interface kinetics in establishing the potential.

2 More recently, Simon et al. [136] have been able to synthesize specific compounds which are used in more selective electrodes for Li^+, Na^+, Ca^{2+} and Ba^{2+} ions. Their structures are shown in Fig. 12.

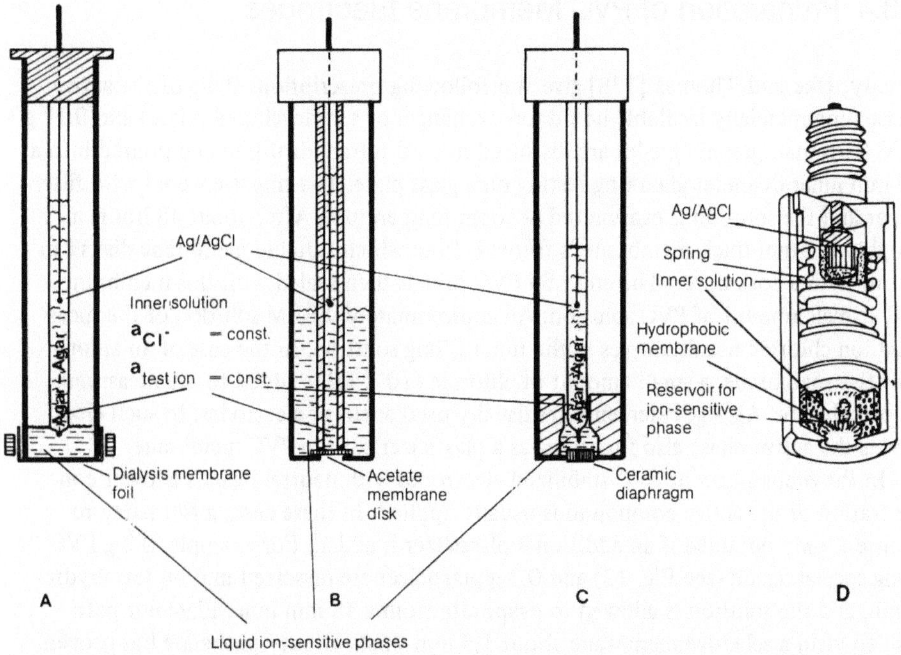

Fig. 28 A–D: Construction of ion-selective electrodes with liquid active phases. **A:** Home-made construction, **B:** Orion liquid exchanger electrodes (Series 92 –), **C:** Corning liquid exchanger electrodes, **D:** Orion (Series 93).

from a reservoir radially surrounding the membrane. The inner side of the membrane is in direct contact with the inner filling solution containing the shunt element. In this construction the thickness of the poorly-conducting exchanger phase is determined by the thickness of the stabilizing membrane. Response times of less than 30 seconds can be attained with this design.

More recently, Orion has introduced series 93, which uses the same basic principles as series 92, but miniaturized to the size of the electrode tip. Its construction is illustrated in Fig. 28. When the electrode stops functioning, after approximately one-half a year's use according to the manufacturer, the electrode tip is simply unscrewed and replaced with a new, factory-tested one. Air bubbles and electric fields interfere far less with this construction than with the series 92.

In the liquid ion-exchanger electrodes of Corning, a ceramic frit imbedded in the electrode body serves as an interface-stabilizer. As shown in Fig. 28, the inner shunt electrode, with its own electrolyte, dips directly into the exchanger phase.

Following the pioneering work of Bloch, Shatkay, Saroff [452] and Moody, Thomas [137], increasingly more manufacturers fix the electro-active compound in a much more sturdy PVC matrix and thus obtain "solid matrix" electrodes with such excellent specifications as stability: < 1 mV/24 h; reproducibility: ≤ 0.5 mV; response time: < 30 s; operational life time: 4-6 months! Because of their importance (some automatic clinical analyzers use them), one way of preparation is given.

3.3.4 Preparation of PVC Membrane Electrodes

Moody, Oke and Thomas [138] give the following prescription: 0.4 g of the active phase (commercially available liquid ion exchanger or self-developed phase) and 0.17 g PVC (chromatographic grade) are dissolved in 6 ml tetrahydrofuran and poured into a 30 mm inner diameter glass ring resting on a glass plate. The ring is covered with filter-paper and the solution is evaporated at room temperature. After about 48 hours a roughly 0.5 mm thick membrane is formed, from which suitable membrane disks can be cut with a cork borer. The end of a PVC tube is then sealed with this membrane and a small amount of PVC solution. An approximately 0.01 M solution of the measured ion chloride usually serves as the inner filling solution. In the case of an anion-sensitive membrane, a small amount of chloride (10^{-3} M) is added to the measured anion solution. Ag/AgCl elements are usually used as shunt electrodes. In such electrodes the active phase also functions as a plasticizer for the PVC membrane.

In the preparation of PVC-stabilized electrodes with neutral ligands a lower concentration of the active compound is usually applied. In these cases a Nernstian response is only obtained if an additional plasticizer is added. For example, 0.2 g PVC, 5 mg special ligand (see Fig. 12) and 0.2 g plasticizer are dissolved in 6 ml tetrahydrofuran, and the solution is allowed to evaporate from a 48 mm inner diameter petri dish to yield a selective membrane about 1.5 mm thick. Dibutylphthalate has proven to be a better plasticizer than tributylphosphate or diphenyl ether [139]. By using less polar plasticizers (such as dipentylphthalate, $\epsilon \approx 5$) shorter response times can be attained [140].

3.3.5 Characteristics

The EMF of a cell consisting of a liquid membrane electrode and a suitable reference electrode dipping in a test solution is given by (compare with Fig. 29):

$$E = E' \pm \frac{R \cdot T}{z_M \cdot F} \ln \left[a_M + \sum_I k_{M-I}^{pot} \cdot a_I^{z_M/z_I}\right]. \qquad (43)$$

The meaning of the symbols is explained in Chap. 1.5. For cations the logarithmic term has a positive sign, for anions it is negative. *In contrast to the homogeneous and heterogeneous precipitate membrane electrodes, whose selectivity coefficients are determined by the corresponding solubility products and are thus independent of the manufacturer, large differences exist among liquid membrane electrodes produced by different manufacturers.* This is due in part to the fact that various different ion-exchanger or ion-solvating compounds are used, and also because different solvents and support materials are employed, all of which can have an influence on the selectivity. For specific applications it is highly advisable to study selectivity coefficients reported by the manufacturers keeping in mind the interfering ions likely to be present. It may well be the case that one firm's electrode exhibits a particularly good selectivity coefficient for a certain interfering ion, while for a different interfering ion the electrode

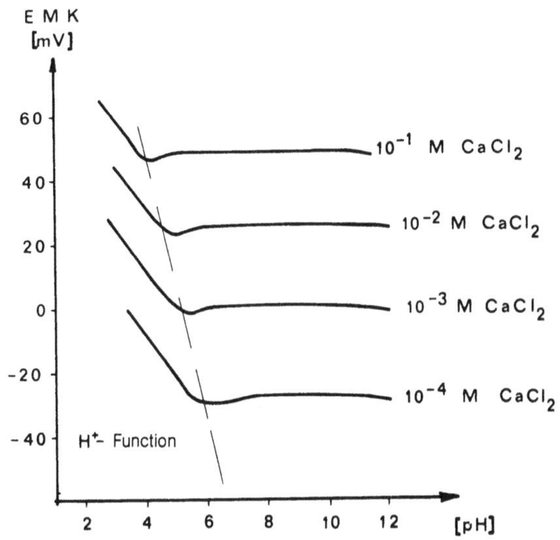

Fig. 29. Influence of pH on the EMF indicated by a pCa ion-exchanger electrode

from a different manufacturer would be better. In critical cases it is a good idea to have the electrode demonstrated before purchase, so that the working range can be determined with prepared standard solutions (see Chap. 5.9). One should also keep in mind that the selectivity coefficients may change with the age of the liquid membrane electrode.

The response time of this type of electrode is longer than those of the electrodes previously discussed because of the smaller exchange current density [453]. In 10^{-3} M solutions it lies between 10 and 30 seconds. This value is, of course, greater near the detection limit and in the presence of interfering ions for the reasons outlined in Chap. 1.4.

Table 4 shows a few additional properties of electrodes of this type.

If the difference in concentration of the measured ion between the inner solution and the sample solution is too large (> 3 or 4 orders of magnitude), then deviations from the Nernst equation can be detected due to a diffusion potential inside the ion-selective membrane [68]. For more accurate measurements in such cases, the concentration of the inner filling solution is adjusted to roughly match that of the sample solution. Most manufacturers use a 10^{-2} to 10^{-3} M inner solution. If the inner solutions is changed, one must remember to add a definite amount of chloride ion for a stable equilibrium Galvani potential at the Ag/AgCl shunt element.

With all liquid membrane electrodes, stabilized or not, the utmost caution is required when working with partially-aqueous media. Due to the solubility of the active phase in such media, the well-defined phase boundary is destroyed, and under some conditions the entire active phase may dissolve out of the electrode.

Lipophilic detergent ions carrying charges of the same sign as the measured ion interfere in all electrodes of this type which depend on specific extraction of an ion from the aqueous phase into the organic phase [141, 142]. The nitrate electrode

Table 4. Liquid and solid-matrix membrane electrodes

Electrode type	Active Phase	Potential-determining Ions	Working range [M]	Selectivity coefficient[1] $k^{pot}_{M\text{-}I}$	Recommended pH-range	Temperature range [°C]	[MΩ] at 25°C	Recommended reference electrode	Manufacturer see also Appendix A.8
pCa, liquid	Ca-salt of dialkyl phosphoric acid in dioctylphenyl-phosphonate	Zn^{2+}, Ca^{2+}, Fe^{2+}, Pb^{2+}	1 to 10^{-5}	Zn^{2+}~3.2; Ca^{2+}~1.0; Fe^{2+}~0.8; Pb^{2+}~0.63; Cu^{2+}~0.27; Ni^{2+}~0.080 Sr^{2+}~0.017; Mg^{2+}~0.014 Ba^{2+}~0.010; Na^{+}~10^{-3} K^{+}~10^{-3};	5 – 9	0 to 50 10 to 60	< 25 < 500	normal Ag/AgCl	Orion Corning
pCa PVC matrix	Ca-dioctyl-phenyl-phosphate	s. above	1 to 10^{-5}	Zn^{2+}~1–5; Ca^{2+}~1.0; Al^{3+}~0.90; Mn^{2+}~0.38; Cu^{2+}~0.070; Fe^{2+}~0.045; Co^{2+}~0.042; Mg^{2+}~0.032; Ba^{2+}~0.020; Na^{+}~10^{-5}; K^{+}~10^{-6}; Li^{+}~10^{-4}	5 to 11	0 to 60	~ 2		Radiometer
	unknown								HNU
pCa PVC matrix	Ca^{2+}-carrier (s. Fig. 12)	Ca^{2+}	1 to 10^{-6}	Sr^{2+}~10^{-2}; Li^{+}~10^{-3}; Cs^{+}~10^{-3}; Na^{+}; K^{+}; Ba^{2+}; Zn^{2+}~10^{-4}; NH_4^{+}~10^{-5}	3 to 10	0 to 50	< 10	s. above	Philips
pMe^{2+} liquid (water hardness)	Ca-salt of dialkyl phosphoric acid in decanol	Zn^{2+}, Fe^{2+}, Cu^{2+}, Ni^{2+}, Ca^{2+}, Mg^{2+}, Ba^{2+}, Sr^{2+}	1 to 10^{-5}	Zn^{2+}~3.5; Fe^{2+}~3.5; Cu^{2+}~3.1; Ni^{2+}~1.35; Ca^{2+}~1.0; Mg^{2+}~1.0; Ba^{2+}~0.94; Sr^{2+}~0.54; Na^{+}, K^{+}~0.01	s. above	0 to 50 10 to 60	< 25 < 500	s. above	Orion Corning
pBa liquid	not given	Ba^{2+}	not given	Ca^{2+}~0.03; K^{+}~10^{-2}; Na^{+}~$4 \cdot 10^{-4}$	not given	10 to 60	< 500	s. above	Corning

[1] according to the manufacturer

Table 4 (continued)

pBa PVC matrix	Ba^{2+}-carrier (s. Fig. 12)	Ba^{2+}	1 to 10^{-6}	$H^+ \sim 0.06$; $Sr^{2+} \sim 0.03$; K^+; $Rb^+ \sim 10^{-2}$; NH_4^+, Cs^+, $Na^+ \sim 3 \cdot 10^{-3}$; Li^+, $Ca^{2+} \sim 10^{-4}$; $Mg^{2+} \sim 10^{-5}$	3 to 11	0 to 50	< 25	s. above	Philips
$pClO_4$ liquid	$Fe(o\text{-}phen)_3^{2+}$ in p-nitrocymene	ClO_4^-, OH^-	0.1 to 10^{-5}	$OH^- \sim 1.0$; $I^- \sim 1.2 \cdot 10^{-2}$; $NO_3^- \sim 1.5 \cdot 10^{-3}$; $Br^- \sim 5.6 \cdot 10^{-4}$; $OAc^- \sim 5.1 \cdot 10^{-4}$; $HCO_3^- \sim 3.5 \cdot 10^{-4}$; $F^- \sim 2.5 \cdot 10^{-4}$; $Cl^- \sim 2.2 \cdot 10^{-4}$; $SO_4^{2-} \sim 1.6 \cdot 10^{-4}$	3 to 10	0 to 50	~ 25	s. above	Orion
pNO_3 liquid	$Ni(o\text{-}phen)_3^{2+}$ in p-nitrocymene tridodecylhexadecylammonium nitrate in n-octyl-2-nitrophenyl ether	ClO_4^-, I^-, ClO_3^-, NO_3^-	1 to 10^{-5}	$ClO_4^- \sim 10^3$; $I^- \sim 20$; $ClO_3^- \sim 2$; $Br^- \sim 0.9$; $S^{2-} \sim 0.57$; $NO_2^- \sim 6 \cdot 10^{-2}$; $CN^- \sim 2 \cdot 10^{-2}$; $HCO_3^- \sim 2 \cdot 10^{-2}$;	3 to 10 1 to 9	0 to 50 10 to 60	~ 25 < 500	s. above	Orion Corning
pNO_3 PVC-matrix	tetradodecylammonium nitrate unknown			$Cl^- \sim 6 \cdot 10^{-3}$; OAc^-, CO_3^{2-}, $S_2O_3^{2-}$, $SO_3^{2-} \sim 6 \cdot 10^{-3}$; $F^- \sim 9 \cdot 10^{-4}$; $SO_4^{2-} \sim 6 \cdot 10^{-4}$; $H_2PO_4^-$, $PO_4^{3-} \sim 3 \cdot 10^{-4}$; $HPO_4^- \sim 8 \cdot 10^{-5}$	3 to 11 1 to 9	5 to 40 0 to 50	~ 25 ~ 25		EIL Philips HNU
pBF_4 liquid	$Ni(o\text{-}phen)_3^{2+}$ in p-nitrocymene	I^-, BF_4^-	0.1 to 10^{-5}	$I^- \sim 20$; $NO_3^- \sim 0.1$; $Br^- \sim 4 \cdot 10^{-2}$; OAc^-, $HCO_3^- \sim 4 \cdot 10^{-3}$; F^-, Cl^-, $SO_4^{2-} \sim 10^{-3}$	2 to 12	0 to 50	~ 25	s. above	Orion

Table 4 (continued)

pLi PVC matrix	Li^+-carrier (s. Fig. 12)	Li^+, H^+	0.1 to 10^{-5}	H^+~1; Na^+, NH_4^+~0.05; K^+~$7\cdot10^{-3}$; Rb^+~$4\cdot10^{-3}$; Cs^+, Ca^{2+}, Sr^{2+}, Mg^{2+}, Ba^{2+}~10^{-4}	3 to 11	0 to 50	~ 25	s. above	Philips
pK liquid	Valinomycin	Cs^+, Rb^+, K^+	1 to 10^{-5}	Cs^+~1.0; NH_4^+~$3\cdot10^{-2}$; H^+~$1\cdot10^{-2}$; Ag^+~$1\cdot10^{-3}$; Na^+~$1\cdot10^{-2}$; Li^+~$1\cdot10^{-4}$;	2 to 11	0 to 50	< 25	Ag/AgCl with double salt bridge (1 M Li-trichloroacetate)	Orion
	tetrakis-(p-chlorophenyl)-borate	Rb^+, K^+, Cs^+,	1 to 10^{-5}	NR_4^+~>1; Rb^+~2; Cs^+~0.4; Cs^+~0.4; NH_4^+~0.3; Na^+, Li^+~10^{-3}; Ca^{2+}, Mg^{2+}~10^{-3}	s. above	15 to 50	~100		Corning
PVC matrix	Valinomycin	Rb^+, K^+, Cs^+	1 to 10^{-6}	Rb^+~1.9; Cs^+~0.38; NH_4^+~10^{-2}; Sr^{2+}~10^{-3}; Li^+, Ba^{2+}~10^{-4}; H^+, Na^+, Ca^{2+}~10^{-5}	s. above	0 to 50	< 2		Philips Radelkis
pNH_4 PVC matrix	Nonactin-Monactin	NH_4^+	1 to 10^{-6}	K^+~0.12; Rb^+~$4.3\cdot10^{-3}$; H^+~$1.6\cdot10^{-2}$; Cs^+~$4.8\cdot10^{-3}$; Li^+~$4.2\cdot10^{-3}$; Na^+~$2\cdot10^{-3}$; Sr^{2+}~10^{-4}; Ba^{2+}, Ca^{2+}~10^{-5}; Mg^{2+}~10^{-5}	4 to 10	0 to 50	< 1	s. above	Philips

[1] according to the manufacturer

can be successfully adapted to indicate MnO_4^-, IO_4^- [143], ReO_4^-, SCN^- [144], phthalate [145] and carboxylate ions. In the last case it has been shown that the selectivity towards carboxylate anions decreases with increasing size of the ion and charge delocalization [146].

3.3.6 Handling

Since all liquid membrane electrodes with inner solutions possess a sensitive inner interface they may only be used in a vertical position with the active phase below. With individual electrodes one should refer to the manufacturer's instructions.

Concerning storage, the reader should be aware that electrodes with stabilized membranes exhibit much shorter lifetimes than the liquid types with reservoirs. Thus it is *not* expedient to store these electrodes in solutions containing the measured ion. They are best stored in a humid atmosphere and conditioned for ½ to 2 hours shortly before the measurement in a solution of the measured ion of concentration comparable to that of the sample solution. In contrast, the liquid membrane electrodes containing a large reservoir of the active phase can be stored in 10^{-3} M solution of the measured ion. *With all electrodes the sensitive electrode surface should never be touched with the fingers.* This can result in a thin layer of oil on the surface which can influence the electrode function. *As soon as the electrode shows a very great (> 20%) deviation from the Nernst equation, the active phase should be replaced. Further attention is required, because small amounts of the active phase can become trapped in the reference electrode diaphragm, leading to a potential term which is also dependent on the activity of the measured ion in solution. In the extreme case the measuring electrode potential will be exactly compensated for (two identical electrodes in a common electrolyte; see Chap. 2.1).*

Liquid membrane electrodes should only be dipped for short periods of time in solutions which contain hydrophobic compounds, such as proteins or oils, in order to avoid a "poisoning" of the membrane. Between measurements in such solutions the electrode should be washed with copious amounts of distilled water. Poisoning of the membrane (for example by picrate anion with the potassium electrode) exhibits itself through a sudden loss of Nernstian response to the measured ion. With PVC-stabilized membranes surface impurities can be removed by wiping with a moist paper towel. Poisoning of electrodes with membrane filters requires replacement of the filter.

If detergent ions with opposing charges are present, then the electrode should be conditioned in the presence of this detergent to avoid potential drift [141].

3.3.7 Further Electro-Active Phases

Looking back at the considerations mentioned in Chap. 3.3.1, a large number of ion-associating molecules can be chosen as the active phase. For testing purposes one need not construct the entire electrode oneself. In many cases one can start with a commercially available liquid membrane electrode (for example Orion, Series 92) and fill it with the organic liquid to be tested. Problems can arise if the viscosity of this new solution is very different from that of the solution normally used with this electrode. In each case a number of membranes having different pore sizes are tested, treating them with dichlorodimethylsilane first to render them hydrophobic. The most well-suited membrane is that which allows the quickest response time while still maintaining Nernstian behavior.

With active phases of intermediate viscosity, a cellulose acetate membrane having 0.1 μm average pore size (millipore VC) has proven most useful [147]. Membrane filters (SM-113-06) of the Sartorius firm (BRD) silanized with trimethylsilyl chloride in benzene are also suitable [148]. A 4 mm diameter disk (appropriate for the Orion Series 92 electrode body) is cut out of the membrane with a cork borer, being careful not to touch it with the fingers. The silyl chloride treatment can be skipped, if, after installing the membrane in the electrode body, one adds the organic phase first and waits until the membrane becomes transparent before adding the inner solution (containing the measured ion in its chloride form, or if the measured ion is an anion containing an additional chloride salt). If membranes from Orion are to be used, they must be washed with petroleum ether and dried before use [149].

The large number of organic compounds with selective solvating properties still to be tested holds a good deal of promise. The following examples should serve to illustrate this point.

When the large tricaprylammonium cation (Aliquat 336 S, General Mill, USA) dissolved in l-decanol (0.01 to 0.1 M) is used, the resulting liquid membrane electrode responds to many anions. In order to obtain stable potentials, the organic phase must be shaken ahead of time with an approximately 1 M solution of the sodium salt of the anion to be measured [150]. In this way, electrodes have been reported in the literature which are selective for HPO_4^{2-} [151], CrO_4^{2-}, $ZnCl_4^{2-}$, $PdCl_4^{2-}$ [152], MnO_4^-, $Cr_2O_7^{2-}$, IO_4^- [143], salicylate$^-$ [153], $HgCl_4^{2-}$, HgI_4^{2-}, $FeCl_4^-$ [154] and CO_3^{2-} [147] anions.

For environmental studies, electrodes which detect *detergents* are of great interest [155]. These also correspond to the same basic type described above: A long-chain alkylpyridinium ion is dissolved in nitrobenzene and shaken with an aqueous solution of the corresponding anionic detergent, such as dodecylsulfonate, tetrapropylenebenzenesulfonate or dioctylsulfosuccinate. A liquid membrane electrode filled with such an active phase allows the determination of the critical micelle concentration [156].

A new type of ion-solvating compound, whose excellent selectivity is perhaps achieved through a helix-forming conformation with the corresponding cation, has been described by Levins [157]. The nonylphenol-adduct of polyethylene glycol and tetraphenylborate dissolved in p-nitroethylbenzene responds selectivity to Ba^{2+} ions in the concentration range 10^{-1} to 10^{-5} M. The selectivity coefficient relative to calcium ion $k^{pot}_{Ba\text{-}Ca}$ is about 10^{-4}. Methylene blue dissolved in o-dichlorobenzene can be utilized as an active phase for the uranyl tribenzoate anion. The corresponding electrode enables in-the-field trace analyses of uranium in rocks [158]. A saturated solution of cesium tetraphenylborate in 4-ethylnitrobenzene can be used as an active phase for a cesium electrode [149], and a solution of 0.2 M thallium(I)-o,o′-dodecyldithiophosphate in chlorocyclohexane as active phase for a thallium electrode [148].

It is up to the interested and innovative experimenter to test other association complexes for their usefulness as specific electrode phases.

3.4 Solid State Electrodes with Electroactive Coatings

The Radiometer Firm (Denmark) offers a universal electrode, the "Ruzicka Selectrode®", whose selectivity can easily be changed. It consists of a pellet of teflon-coated, hydrophobic graphite powder surrounded by a suitable teflon shell. A simple rust-proof wire is fed into the graphite as a shunt electrode [159]. Such graphite electrodes were previously used exclusively for voltammetric investigations. If an electroactive substance (for example, the various silver halides, Ag_2S-metal sulfide mixtures, etc.) is rubbed into the graphite surface (a rounded glass rod is used for this), the corresponding ion-selective electrode results (compare with Chap. 3.2.2). Polishing the active surface enhances the response time and lessens the surface pore memory effects. With a slight modification the liquid active phases can also be fixed onto the graphite surface. The selectivity of this selectrode corresponds to that of the other constructions. Since there is no well-defined potential shunt system in this electrode, the E' term in the Nernst equation can fluctuate. This can be compensated for, however, by frequent calibration. If one wants to switch from one measured ion to another, one simply cuts a thin slice of the electrode tip off, with the help of the cutting knife supplied, and activates the freshly exposed graphite surface with the active phase selective for the desired ion.

A similar construction has been described by Ansaldi and Epstein [160]. They fixed a 0.5 mm thick calcium-selective PVC membrane (preparation: see Sec. 3.3.4) on the surface of a graphite electrode (used for spectroscopy), and obtained a solid-state calcium electrode which exhibited the characteristics of the fluid membrane type.

Simon, Morf and Pretsch [161] discuss solid-state microelectrodes consisting of ion-selective PVC membranes for Ca^{2+}, Ba^{2+}, Li^+, Na^+ and K^+ mounted on a silver surface which has been pre-treated with tetraphenylborate. These electrodes have lifetimes of a year and more. The EMF stabilities were excellent, and lacking a fluid/fluid phase boundary, the electrodes were less pressure-sensitive.

James, Garmack and Freiser [162] described a wire electrode construction for fluid exchanger systems which contained no inner solution. These electrodes are easy to produce and to miniaturize. For this reason, and also because they are selective for certain amino-acids, these electrodes are especially interesting for physiological measurements. Since they are not yet commercially available, extensive practical experiences are lacking. The active phase in this construction is a liquid organic ion-exchanger containing the large quarternary tricaprylmethylammonium cation (Aliquat 336 S, General Mill, USA; Serva, Heidelberg). These electrodes respond to anions, so the detection of amino-acids requires high pH values to insure that the amino-acids are predominately in their anionic form.

Before measuring any of the ions listed in Table 5, the organic exchanger phase must first be saturated with the appropriate measured ion, in order to achieve sufficient electrode stability (see the discussion of electrode phase identity in Chap. 1.3). To accomplish this, a 0.5 to 1 M aqueous solution of the sodium salt of the measured ion is shaken with 60 vol. % of the tricaprylmethylammonium compound in l-decanol. If an emulsion results, it is centrifuged. This shaking procedure is repeated until no more chloride is to be found in the aqueous washings (Aliquat 336 S is usually supplied in its chloride form). This usually takes about six washings. For leucine and phenylalanine a 30 vol. % solution of the quarternary ammonium salt is recommended. The organic phase resulting after successive washings represents the active phase. To prepare a suitable support, a thin, recently annealed platinum wire is covered with a layer of polyvinylchloride. This can be done by dipping the wire in a 15% solution of PVC (Serva, Heidelberg) in cyclohexanone and allowing it to dry (~ 30 minutes). The PVC layer is then "steeped" in the active phase for two hours. Alternatively, these two steps can be combined by dipping the platinum wire directly into a 10:1 solution of 8% PVC solution/active phase, and allowing it to dry for about 1 hour. In both cases the uncovered portion of the platinum wire must be covered with paraffin to avoid direct contact with the test solution.

Many of the ions listed in Table 5 are indicated by the corresponding coated wire electrodes exactly according to the Nernst equation in the concentration range 10^{-1} to 10^{-4} M. For sulfate and oxalate anions a slope of about 28 mV per power of ten in activity results at 25°C. Table 5 illustrates the good selectivity behavior of these electrodes. In all cases the selectivity coefficients are more favorable than those for the corresponding liquid membrane electrodes [150]. The response time is very short – only a few seconds. The lifetime of these microelectrodes is claimed to exceed 3 months.

In the interest of avoiding potential drift, these ion-selective coated wire electrodes should be conditioned in a solution about 0.1 M in the measured ion for about 15 minutes prior to the first use. Between measurements they can be kept in air, and after longer periods of non-use condition again in a 0.1 M solution for about 5 minutes. Care must be taken that no direct electrolytic contact is made between the platinum wire and the sample solution, since naturally this would give false readout.

The coated-wire electrodes are very interesting from a theoretical standpoint. Different selectivities can be attained with the same active phase by changing only the conditioning. Since only the concentration of the measured ion in the active phase is increased during conditioning, the resulting selectivity can be explained in that this rise in concentration also increases the corresponding exchange current density. Also, the appearance of a mixed potential can be detected with these electrodes, since a few of the detergent electrodes exhibit a slope which exceeds the theoretical slope of the Nernst equation.

Finally, it should be mentioned that with these electrodes it is difficult to reproduce the results of other workers. In some cases long lasting potential drifts arise, in others the active compound dissolves out of the PVC matrix. Because of the undefined and unknown potential-creating process at the metal/PVC phase boundary, one should work with electrometer amplifiers showing the highest input impedance ($> 10^{14}\ \Omega$) and lowest current to avoid any polarization of the unknown redox reaction at the metal/PVC interface.

Table 5. Coated wire electrodes [162–164]

Electrode type	Active Phase	Potential-determining Ions	Working range [M]	Selectivity coefficient $k^{pot}_{M\text{-}I}$	Recommended pH-range	Temperature [°C]	[MΩ] at 25°C	Recommended Reference Electrode
pCl	Aliquat 336S in PVC-Matrix	NO_3^-, Br^-, Cl^-	10^{-1} to 10^{-4}	NO_3^-~2.0; Br^-~1.2; SO_4^{2-}~0.12	2–12	0 to 50	~100	Ag/AgCl with double salt bridge
pBr	s. above	I^-, NO_3^-	10^{-1} to 10^{-3}	I^-~14.5; NO_3^-~2.0; Cl^-~0.19; SO_4^{2-}~0.020	s. above			
pI	s. above	I^-	10^{-1} to 10^{-4}	NO_3^-~0.11; Br^-~0.056; Cl^-~0.0048; SO_4^{2-}~0.001;	s. above			
pSCN	s. above	SCN^-, I^-	10^{-1} to 10^{-3}	I^-~0.34; NO_3^-~<0.046; SO_4^{2-}~0.001; Cl^-~< 0.001	s. above			
$pClO_4$	s. above	ClO_4^-	10^{-1} to 10^{-4}	ClO_3^-~0.039; Cl^-~0.004; NO_3^-~0.028; SO_4^{2-}~< 0.001	s. above			
pSO_4	s. above	NO_3^-, Cl^-, SO_4^{2-}	10^{-1} to 10^{-3}	NO_3^-~30; Cl^-~16	s. above			
p-Oxalate	s. above	Cl^-, OAc^-, Oxalate$^-$	10^{-1} to 10^{-4}	Cl^-~51; OAc^-~11; SO_4^{2-}~1.3	> 8			
p-Benzoate	s. above	Salicylate$^-$, Cl^-, Benzoate$^-$	10^{-1} to 10^{-3}	Sal^-~1.4; Cl^-~1.3; NO_3^-~1.3; SO_4^{2-}~0.15	>10			
p-Salicylate	s. above	Sal^-, NO_3^-	10^{-1} to 10^{-3}	NO_3^-~0.42; m–OH-benzoat$^-$~0.22; Cl^-, SO_4^{2-}~< 0.001;	>10			
p-Leucine	s. above	Phe^-, Leu^-, Cl^-, NO_3^-	10^{-1} to 10^{-3}	Phe^-~2.0; Cl^-~1.3; NO_3^-~0.50; Val^-~0.25 Gly^-~0.032; SO_4^{2-}~0.025	>10			

Table 5 (continued)

Electrode Type	Active Phase	Potential-determining Ions	Working range [M]	Selectivity coefficient k_{M-I}^{pot}	Recommended pH-range	Temperature [°C]	[MΩ] at 25°C	Recommended Reference Electrode
p-Phenylalanine	s. above	NO_3^-, Phe^-, Cl^-	10^{-1} to 10^{-3}	$NO_3^- \sim 2.0$; $Cl^- \sim 0.8$; $Leu^- \sim 0.13$; $SO_4^{2-} \sim 0.020$ $Gly^- \sim 0.025$	>10			
p-$FeCl_4$	s. above	$FeCl_4^-$, $FeCl_4^{2-}$	10^{-1} to 10^{-4}	$Fe^{2+} \sim 0.1$; $Hg^{2+} \sim 10^{-2}$; $Sn^{2+} \sim 0.05$; $Zn^{2+} \sim 0.05$ $SO_4^{2-} < F^- < NO_3^- \sim 10^{-2}$	0 (1 M HCl + 5 M LiCl)			
p-Toluenesulfonate	s. above	p-$CH_3C_6H_4SO_3^-$	10^{-1} to 10^{-5}	$Cl^- \sim 10^{-2}$; $SO_4^{2-} \sim 10^{-2}$; $NO_3^- \sim 1$; $ClO_4^- \sim 1$; $Ac^- \sim 0.5$	>10			
p-Laurylbenzenesulfonate	s. above	$C_{12}H_{25}C_6H_4SO_3^-$	10^{-3} to 10^{-6}	as above	>10			
p-Laurylsulfate	s. above	$C_{12}H_{25}SO_4^-$	10^{-2} to 10^{-4}	as above	>10			

3.5 Gas Sensors For CO_2, NH_3, SO_2, NO_2, HF, H_2S, HCN, etc.

Although these simple compounds are all neutral gases, they will be briefly discussed here because, in addition to the pH glass electrode, in some cases (HF, H_2S and HCN) an ion-selective electrode is employed in the construction of the relevant sensor.

3.5.1 Principles

Common to all these systems is the reaction of the respective gas with water to produce ionic species:

$$CO_2 + H_2O \rightleftharpoons HCO_3^- + H^+ \qquad (R27)$$

$$NH_3 + H_2O \rightleftharpoons NH_4^+ + OH^- \qquad (R28)$$

$$SO_2 + H_2O \rightleftharpoons HSO_3^- + H^+ \qquad (R29)$$

$$2\,NO_2 + H_2O \rightleftharpoons NO_3^- + NO_2^- + 2\,H^+ \qquad (R30)$$

$$HF + H_2O \rightleftharpoons H_3O^+ + F^- \qquad (R31)$$

$$H_2S + H_2O \rightleftharpoons HS^- + H_3O^+ \qquad (R32)$$

$$HCN + H_2O \rightleftharpoons H_3O^+ + CN^-. \qquad (R33)$$

In the cases where hydrogen ions are released or consumed upon introduction of the gas into the aqueous phase, pH glass electrodes can be used to indicate the change in pH of the test solution. In the last three cases, ion-selective electrodes (fluoride and silver-selective electrodes) can be employed.

According to the corresponding mass action expressions,

$$K = \frac{a_{HCO_3^-} \cdot a_{H^+}}{a_{CO_2}}, \quad a_{H^+} = K \cdot \frac{a_{CO_2}}{a_{HCO_3^-}} \qquad (44)$$

$$K = \frac{a_{NH_4^+} \cdot a_{OH^-}}{a_{NH_3}}, \quad a_{OH^-} = K \cdot \frac{a_{NH_3}}{a_{NH_4^+}}, \text{ etc.} \qquad (45)$$

at a constant (buffered) HCO_3^-, NH_4^+, etc. activity a direct proportionality exists between the concentration of the neutral gas and that of the electrochemically indicated ion.

3.5.2 Construction

The above measuring principle can be put into practice with two different types of constructions. One construction follows the example of the CO_2 electrode [165, 166], which has been commercially available for a long time and is most frequently used for clinical pCO_2 measurements. Here the needed reaction solution is held in a thin layer before the surface of the indicating electrode with the help of a gas-permeable membrane and a spacer ring. Since the extremely thin membrane lets only gases pass through, no interfering ions can penetrate to the electrode. The other construction uses a small air space (air-gap-electrode [167]) to keep the ionic constituents of the sample solution away from the electrode surface. Here the indicator electrode surface must be coated with the corresponding reaction solution before each measurement (accomplished by contacting the surface with an electrolyte-soaked polyurethane foam). Figure 30 illustrates both construction designs. Since combination electrodes (with reference elec-

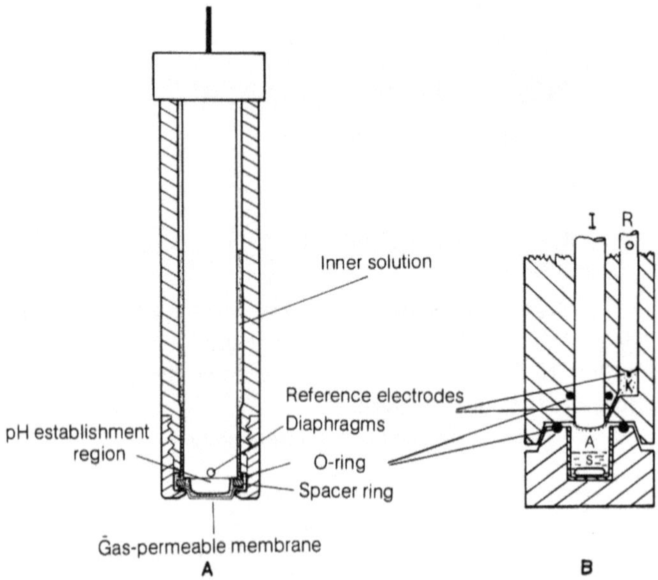

Fig. 30. Construction of gas-sensitive electrodes. **A:** Membrane type, **B:** Airgap type (A = air space, S = stirrer)

trode incorporated) with flat electrode tips are used as indicator electrodes in the membrane covered type, an additional, outer reference electrode is unnecessary. One advantage of the air-gap electrode is its quick response time. The diffusion of the gas through the 10 mm air space is faster than the diffusion through the micropores in the gas-permeable membrane. In addition, in contrast to the membrane-covered type, equilibrium is reached sooner in the extremely thin, limited volume electrolyte layer of the air-gap-electrode. Since this electrode never comes in direct contact with the test solution, all the membrane-related interferences are avoided (plugging of the pores by viscous media, protein layers, etc., penetration of the sample solution due to the presence of a wetting agent, etc.). These advantages are paid for by a somewhat troublesome handling procedure, since the coating on the electrode surface must be reproducibly applied!

3.5.3 Characteristics of Gas-Sensitive Electrodes

Table 6 summarizes the most important features of the commercially available gas-sensitive electrodes. In general, the gas transport through the membrane foil (as through the air-gap) is diffusion controlled. The driving force for this diffusion is the partial pressure gradient. Thus when a gas-sensitive electrode is dipped into a solution it indicates the different partial pressures of the gases in solution. The relationship between the partial pressure of a gas P_g and the concentration of physically dissolved material [C] is described by Henry's Law:

$$P_g = k \cdot [C]. \tag{46}$$

The proportionality constant k reflects the physical solubility of the corresponding material in the solution. It is dependent upon the composition of the solution. *Because k should be the same for the calibration solution and the sample solution, the calibration solution should resemble the sample solution as closely as possible.* In cases where the composition of the sample solution is in doubt, it should be diluted and remeasured. According to the same principle, water vapor transport can also occur if the osmotic pressures of the sample and reaction solutions differ. This process can dilute the concentrations in the thin electrolyte film and cause the measured values to drift.

Isotonic solutions can be attained if the ionic strengths of the sample solution and the reaction solution are the same. Many gas-sensitive electrodes closely follow the Nernst equation between 0.1 and 10^{-5} M:

$$E = E' \pm \frac{R \cdot T}{z \cdot F} \cdot \ln\,[\text{gas}]. \tag{47}$$

Caution is required when working in strongly basic solutions (pH > 13), *because upon long exposure to such media the membrane may swell.* This can decrease the size of the pores necessary for gas transport and influence the diffusion. The same applies to working in non-aqueous solvents. In order to achieve the full sensitivity of these gas sensors, it is recommended that one work with sample solutions within the optimal pH ranges listed in Table 6. In the absence of ammonia, the NH_3-sensor can be used to indicate volatile amines (for example, to monitor the air in the vicinity of a plant manufacturing such a substance). To free NH_3 bound in metal complexes (copper or zinc tetrammine complexes, for example) the addition of 15 g/l EDTA is recommended [168].

The optimal pH value for SO_2 and NO_2 is below 0.7, and can be established with H_2SO_4 or $HClO_4$. If one is forced to use pH's at which the equilibrium does not lie

Table 6. Gas sensors [167, 170]

Indicated Gas	Indicator-electrode	Membrane	Reaction solution [M] [1] [3]	Detection limit [M]	optimum pH-value	Interferences	Manufacturer see also Appendix A.8
CO_2	pH-Glass electrode	micropore (1,5 μ) filter [2]	10^{-2} $NaHCO_3$ 10^{-2} NaCl	$\sim 10^{-5}$	< pH 4		Ingold Radiometer HNU
NH_3	s. above	0.1 mm micropore Teflon	10^{-2} NH_4Cl 0.1 KNO_3	$\sim 10^{-6}$	> pH 12	volatile amines	Beckman EIL Orion Radelkis HNU
SO_2	s. above	0.025 mm Silicon rubber	10^{-3} $NaHSO_3$ pH 5	$\sim 5 \cdot 10^{-6}$	< pH 0.7	destroy Cl_2, NO_2 with N_2H_4; HCl, HF, acetate	EIL
NO/NO_2	s. above	0.025 mm micropore polypropylene	0.02 $NaNO_2$ 0.1 KNO_3	$\sim 10^{-6}$	< pH 0.7	remove SO_2 with CrO_4^{2-}; CO_2	EIL Orion HNU
HF	Fluoride electrode	micropore filter [2]	1 H^+	$\sim 10^{-3}$	< pH 2		Orion
H_2S	Ag_2S-Membrane electrode	s. above	Citrate buffer pH 5	$\sim 10^{-8}$	< pH 5	reduce O_2 with ascorbic acid	Orion
HCN	s. above	s. above	10^{-2} $KAg(CN)_2$	$\sim 10^{-7}$	< pH 7	remove H_2S with Pb^{2+}	Orion

[1]) With air-gap electrodes an additional 0.1% of a non-ionic wetting agent is added (such as Victarwet (Stauffer Chemical Cor.) [167])

[2]) such as Fluoropore with 1 μ pore size and a polyethylene backing (Millipore) [171]

[3]) membrane-covered sensors may also contain an agent to increase the viscosity [172]

entirely on the side of the gas, a reduced sensitivity results. In order to get comparable results in such cases, the pH of each sample solution must be adjusted to exactly the same value! Near the detection limit, the response time may vary between 10 sec and a few minutes, depending on the contents of the sample solution. The evaluation methods and the actual measuring techniques are similar to other ion-selective electrodes, so the points discussed previously in these connections apply here as well [169].

3.5.4 Handling

All membrane-containing gas sensors should be stored in a solution comparable to the inner solution when not in use. The electrolyte solutions listed in Table 5 can be used for this. Prolonged storage in air will cause the inner solution to become more concentrated, possibly leading to precipitation of salts on the inner side of the membrane. After one week to one month, both the membrane and the inner solution should be replaced. In detecting traces of a gas, a diluted reaction solution can be advantageous. *The membrane itself should never be touched with bare fingers.* Air bubbles in the reaction phase are also to be avoided. One should also be aware of the fact that the gases set free by the optimal pH will escape when not actually being measured by the air-gap-electrode. Thus the establishment of the proper pH should only be carried out immediately prior to the measurement. *The volume to surface area ratio of the sample solution should be as large as possible. With stirred solutions 50% of the dissolved gas is lost in about 6 hours.* In the cases where a pH glass electrode serves as indicator electrode, the response time can grow longer after a few months due to changes in the glass membrane. Reactivation is possible by alternately treating the glass electrode with 0.1 M HCl and 0.1 M NaOH. In general, follow closely the directions of the manufacturer.

3.5.5 Sample Preparation

The points which must be observed for successful operation of these gas sensors are collected below:

a) Calibration and sample solutions should not differ greatly in composition (same proportionality factor in Henry's Law).

b) Calibration and sample solutions should be measured at the same temperature (otherwise the proportionality factor changes).

c) The optimal pH value (see Table 5) should only be established immediately prior to the measurement.

d) The ionic strengths of the sample and reaction solution should be similar (otherwise water vapor transfer and drifting values result).

e) In all cases vigorous stirring is necessary. If the measured gas can escape, then a closed system must be used.

f) With concentrations above 0.1 M, the saturated region may be reached, in which Henry's Law no longer holds.

g) The accuracy depends on the standard solutions used. Since a $NaHSO_3$ solution (SO_2 electrode) quickly decomposes, an iodometric determination of the solution concentration is used as a check.

In addition to NH_3 and NH_4^+ ions, the NH_3 electrode can also indicate NO_3^-, NO_2^- and *organic nitrogen-containing compounds* after conversion to NH_3. For organically bound nitrogen, the well-known Kjeldahl decomposition is suitable. Nitrate and nitrite can be reduced to NH_4^+ in an HCl – NaF solution containing Al powder, and subsequently converted into NH_3 with NaOH. The yield of this reduction is greater than 90%. *Direct reduction in alkaline solution results in significant loss of gaseous NH_3.* Here also, calibration and sample solutions should be subjected to the same conditions. Since large amounts of chloride interfere with the directly-indicating nitrate electrode, the NH_3 electrode furnishes the ability to test chloride-rich samples for their nitrate content (such as soil or seawater samples), opening a broad new field of applications [169]. The membrane-covered NH_3 electrode also has wide applicability in the field of enzyme electrodes to be discussed in the next chapter. Ruzicka and Hansen [170] reported the successful application of the air-gap NH_3 electrode to the determination of NH_4^+ in serum and plasma samples. With the help of a Technicon Autoanalyzer®, 60 determinations per hour were possible. With the CO_2 electrode, the total inorganic and organic carbon content of water samples could be easily and inexpensively obtained [170].

The SO_2 electrode enables the determination of sulfurous acids in foodstuffs and beverages. Simply acidifying the sample solution to $pH < 0.7$ with $HClO_4$ allows the "free SO_2" to be determined. The aldehyde-bound SO_2 can be released by making the solution more alkaline than pH 12.5, and then re-acidifying to $pH < 0.7$. The recombination reaction is kinetically retarded [168].

3.6 Bio-Sensors

With the exception of the Corning-476200 acetylcholine bromide-selective liquid membrane electrode, there are no commercially available "bio-sensors" which use ion-selective electrodes to indicate biochemically important compounds (such as proteins, antigens, hormones, enzymes and their substrates). Nonetheless, there is a large number of such devices which can be easily self-made, so that it is appropriate to briefly discuss them here. For further details the reader is referred to the excellent review articles by Rechnitz [173] and Moody and Thomas [174].

3.6.1 Principles

In the same manner as the gas-sensors, these electrodes are based on a chemical reaction which consumes or releases some material which can be detected with an ion-selective electrode. The prime example for the so-called *enzyme-electrodes* is the cleavage of urea into ammonium and carbonate ions by the urease enzyme:

$$CO(NH_2)_2 + 2\,H_2O \xrightarrow[\text{(enzyme)}]{\text{urease}} 2\,NH_4^+ + CO_3^{2-} \qquad \text{(R34)}$$
(substrate) (products)

There are five possibilities for monitoring this particular reaction with the help of ion-selective electrodes: The ammonium ions released can be detected by means of a cation-selective glass membrane electrode or an ammonium-selective liquid membrane electrode, or the NH_3 gas sensor can be used following treatment with base. Alternatively, the carbonate ion can be detected either directly with a carbonate-sensitive liquid membrane electrode [146] or after conversion into CO_2 with a CO_2-sensor.

Utilizing reactions such as (R34), not only can the substrate concentration be determined with the help of ion-selective electrodes, but by using an excess of the substrate, the kinetics of the reaction (EMF vs. time curve) and, thus, the enzyme activity can be investigated. In addition, using comparative activity measurements, activators (such as Na-glycolate for lipase) and inhibitors (phosphorus insecticides for cholinesterase) can also be determined. Table 7 lists a few examples.

3.6.2 Construction

In the simplest case, the ion-selective measuring cell can be dipped directly into a stirred, thermostated solution of the substrate, the enzyme added, and a measurement taken after the new potential has been established [175]. The change in concentration of the indicated ion, and thus the concentration of the substrate can be calculated from the change in potential. It is assumed here, naturally, that the reaction is stoichiometric, i.e. the substrate molecules must react quantitatively. The drawbacks of the "beaker method" are the relatively large amount of time required, and the fact that the added enzyme, sometimes quite expensive, must be thrown away, even though it is still active.

Based on the work of Clark and Lyons [176] as well as Hicks and Updike [177], Guilbault and Montalvo [178–181] immobilized the enzyme. They coated a traditional cation-selective glass electrode (responding to NH_4^+) with the enzyme (urease, amino-acid oxidase) in a polymerized acrylamide gel matrix. For better mechanical stability, the gel was reinforced with a nylon net or a cellophane foil. The enzymes reacted specifically only with the urea and amino acids, respectively, in the solution to produce NH_4^+ ions, which were indicated by the cation-selective electrode. This electrode construction functioned as a specific detector for over three weeks with no loss in activity. The urea sensor spanned a concentration range of 1.6×10^{-1} to 5×10^{-5} M with a response time of only approximately 25 seconds.

3.6.3 Preparation of Enzyme Electrodes

a) *Urea Sensor by the Bound Method* [182]
600 I.U. urease is dissolved in 1 ml phosphate buffer (pH 6.8), 1 ml of a 17.5% bovine albumin solution and 0.07 ml of a 25% glutaraldehyde solution are added, and the mixture is stirred for 2 minutes. The cation-selective electrode is washed with water, care-

Table 7. Examples of Bio-sensors

Substrate (substance to be determined)	biologically active compound (activity to be determined)	Product (indicated material)	optimal Buffer and pH-value	Indicator-electrode	Working range	Literature
Acetylcholine (indirect: organic phosphorous pesticides in the ng/ml-range)	Acetylcholinesterase	Choline, acetic acid	physiol. salt soln; pH 7.2	Acetylcholine liquid membrane	10^{-2} - 10^{-4} M	[192, 193, 194, 195]
5'-Adenosinemono-phosphate (5'-AMP)	AMP-Deaminase	5'-Inosinmono-phosphate, NH_3	0.05 M Tris; pH 7.5	NH_3 Gas-Sensor	10^{-2} - 10^{-4} M	[215]
Albumin (Human)	Anti-Human-Serum-Albumin (Immuno-reaction)	unreacted Antibodies	0.1 M NaOH	Ag_2S	0.5-30 μg/ml	[196]
Amygdalin	-Glucosidase	Glucose, Benzaldehyde, CN^-	NaH_2PO_4; pH 6.4	Cyanide	10^{-2} - 10^{-5} M	[197, 198]
L-Arginine	Arginase+Urease (double reaction)	NH_3, CO_2	Tris; pH 7	NH_3-air-gap		[199]
Asparagine	Asparaginase	Aspartic acid, NH_4^+	Tris; pH 7	NH_4^+-Glass membrane		[200]
Catechol	Polyphenoloxidase	o-Quinone	Acetate; pH 5.9	Copper		[174]
Cholesterol	Cholesterolesterhydrolase + Cholesteroloxidase (double reaction)	Cholest-4-en-3-one, H_2O_2	Na_2HPO_4/ NaH_2PO_4; pH 6.8	Iodide	50-400 mg%	[187]
Cyanide/Thiosulfate	Rhodanase	SCN^-	NaH_2PO_4; pH 7.9	SCN or CN		[201]
L-Cystein/Cyanide	-Cyanoalanine-synthase	-Cyanoalanine, HS^-	Tris-Acetate; pH 8.5	Ag_2S		[202]
Diphenylcarbamylfluoride	-Chymotrypsin	DPC-chymotrypsin, F^-	pH 7.5	Fluoride	0.03 μM	[203]
Glucose	Glucoseoxidase	Gluconic acid H_2O_2	KH_2PO_4/K_2HPO_4; pH 5.7	Iodide	1 - 300 mg%	[204]

Table 7 (continued)

Substrate (substance to be determined)	biologically active compound (activity to be determined)	Product (indicated material	optimal Buffer and pH-value	Indicator-electrode	Working range	Literature
Glutamine	Glutaminase	Glutamic acid, NH_4^+	Tris; pH 7	NH_4^+-Glass membrane		[200]
Urea	Urease	NH_3, CO_2	Phosphate; pH 7 - 8.5	NH_4^+-Glass membrane NH_3, CO_2-Gas-Sensor		[181, 183, 184, 190, 205, 206]
Yeast	Concanvalin A (Immuno-electrode)	–	Phthalate; pH 3.5	PVC-wire		[189]
Creatinine	Creatininase	N-Methylhydantoin, NH_4^+	Tris/NaH_2PO_4; pH 8.5	NH_3Gas-Sensor	1 - 100 mg%	[207]
L-Leucine	L-Amino acid oxidase	$RCOCOO^-$, NH_4^+, H_2O_2	Tris; pH 7	NH_4^+-Glass membrane		[200]
D-Methionine	D-Amino acid oxidase	$RCOCOO^-$, NH_4^+, H_2O_2	Tris; pH 7	NH_4^+-Glass membrane		[200]
Nitrite	Nitrite reductase Methylviologen	NH_4^+	Phosphate; pH 7.2	NH_4^+-liquid	10^{-3} M	[208, 209]
Penicillin	Penicillinase	corresp. acid	pH 6.9	pH-Glass membrane	3 - 1100 μg/ml	[210]
L-Phenylalanine	L-Amino acid oxidase	$RCOCOO^-$, NH_4^+, H_2O_2		Iodide		[211]
Sulfate	Sulfate reductase	HS^-		Ag_2S		[212]
D,L-Tyrosine	D,L-Amino acid oxidase	$RCOCOO^-$, NH_4^+, H_2O_2	Tris; pH 7	NH_4^+-Glass membrane		[200]
Tyrosine	Tyrosinedecarboxilase	Tyramine, CO_2	Na-citrate; pH 5.5	CO_2-Sensor		[206]

fully dried with a soft paper towel and dipped in the above solution for a short time. Finally, it is rotated at an angle of 45° for 15 minutes so that a uniform film thickness is obtained during the cross-linking process. In order that the cross-linked layer not become detached from the electrode, a suitable O-ring is used to fasten this layer down above the ion-selective hemisphere. The sensor is then washed with distilled water, glycine, and again with water to wash away any of the excess reagents.

The proper cross-linking time is essential for the electrode function. Too short a time yields a membrane which is too soft and easily torn. Too long a time results in too low an adhesiveness and a severely reduced enzyme activity. With the proper cross-linking period less than 30% of the original enzyme activity is lost.

b) *Urea Sensor by the Sandwich Construction*
60 I.U. urease are dissolved in 0.56 ml phosphate buffer (pH 6.8), 0.4 ml of a 1.5% bovine albumin solution and 0.04 ml of a 25% glutaraldehyde solution are added, and the resulting mixture is poured onto a glass plate. The cross-linking is carried to completion at 4°C. The membrane thus produced is washed with water and removed from the glass plate. A disk is cut out with a cork borer and covered in open-faced sandwich fashion with a gas-permeable membrane of a CO_2 sensor [182].

According to another procedure, a drop of the urease-buffer mixture (pH 7.0) is placed on the outer surface of an NH_3 sensor gas-permeable membrane, covered with a suitable cellulose membrane, and incorporated into the sensor body according to the instructions of the manufacturer [183]. Even though the pH value is such that NH_4^+ ions predominate, the sensor still shows a Nernstian response to the urea concentration. Even full blood analyses are possible.

c) *Urea Sensor Using an Air-Gap Electrode* [184]
10 I.U. of immobilized urease (Boehringer-Enzygel-Urease-650 units/g; Corning glass-bound urease-300 units/g) is coated onto the teflon stirrer of an air-gap electrode, and a nylon net is placed over it and tied at the end. Thanks to the favorable hydrodynamics of this construction, an analysis requires only about two minutes, even though the 2-amino-2-hydroxymethyl-1,3-propandiol buffer used (0.5 M, pH 8.5) results in a pH value at which the enzyme cannot display its optimal activity.

These three types of constructions are only basic examples. The innumerable highly selective enzymes or complete enzyme systems known can open further possibilities.

3.6.4 Characteristics of Bio-Sensors

Only a few general comments concerning the properties of these sensors can be made here, because the properties of the biologically active phases serving as a basis in each individual case (enzyme, antibody, hormone receptor, etc.) play a central role.

In general, the bio-sensors of the above sandwich construction type exhibit a nearly Nernstian relationship between the substrate concentration and the EMF. The response times of constructions of this sort are influenced by many different factors [182]: The enzyme reaction itself procedes at a finite rate which depends upon the initial substrate concentration and the Michaelis constant. The product of this reaction must then diffuse to and from the surface of the indicator electrode as its concentration changes.

The concentration of the indicated material (substrate or product) at the indicator electrode surface under steady state conditions depends on the Michaelis constants, the activity of the biologically active compound in the membrane, the thickness of this membrane, and the diffusion coefficients of the substrate and product. It can be shown that for any given system, the response time of the enzyme electrode is given by the ratio $d^2/\bar{D}$ (d = layer thickness; $\bar{D}$ = effective diffusion coefficient), except if the rate-determining step is the diffusion of a gas through a gas-permeable membrane, as in the case of the membrane-covered gas sensors. In practice layer thickness as low as 30 μ can be used. Depending on the substrate concentration, response times between 10 seconds and 15 minutes can be achieved.

The lifetime of such sensors depends on the stability of the biologically active compound. With sensitive compounds the activity can only be maintained for a few hours. With suitable immobilization and storage in a refrigerator during non-use, the urea sensor, for example, can exhibit an active lifetime of 60 days [184]. The loss of activity can have many causes. Certain heavy metal ions present in trace amounts can act as inhibitors. As an example, it has been reported [185] that an anti-enzyme factor contaminating the solution via the KCl solution of a calomel reference electrode (presumably Hg^+ ions) could be removed with 10^{-3} M cystein. Sometimes the interfering heavy metal ions (Cu^{2+}, Zn^{2+}, etc.) can be bound by adding a complexing agent (EDTA) [186]. It is more difficult to remove interfering anions (such as the nitrate anion with urease). On the other hand, this allows one to indirectly determine the inhibitor concentration [454]. The lifetime of enzyme electrodes can also be shortened by bacterial degradation. To suppress this, 10^{-3} M sodium azide can be added [187, 188]. Certain activators (such as phosphate with ureas, sodium cholate with cholesterolesterhydrolase) can effectively prolong the lifetime of these electrodes.

Since bio-sensors based on enzymes are simple and easy to use, they are already in use in the Technicon Autoanalyzer® for clinical work, and show excellent agreement with analyses obtained with other, more troublesome methods. Bio-sensors based on *antibodies* (immuno-electrodes) are already in the testing phase [173, 189], and those based on *hormones* are in the stage of theoretical conception [173].

3.6.5 Sample Preparation with Bio-Sensors

Here again, it is difficult to give detailed procedures, because each of the large number of possible reactions has its own set of optimal parameters. One problem which often arises is the proper pH value. The optimal pH value for the biologically active material does not always coincide with that of the ion-selective electrode. This may be because H^+ ions are directly sensed by a cation-sensitive glass electrode, or because the species to be indicated is associated (for example, NH_4^+ ion formation when NH_3 is to be detected, or HCN formation when measuring CN^-). In the case of urease, a compromise can be reached between pH 7 [183] and 8.5 [184]. When this is not possible, the reaction and indication must be broken down into two separate steps. This was done in the first attempts to measure urea in whole blood [190]. In such cases it is best to use a flow cell apparatus in which the biological reaction is separated from the indicator electrode, and the optimal indicator electrode pH is established immediately before this electrode is encountered. In many cases the biologically active material can be bound to the inner surface of a nylon tube [191]. Thus a reactor for the Technicon Autoanalyzer® is constructed in this simple, elegant way. The flow rate and tube length (reaction time) are easy to empirically optimize. With flow cell construction there is the additional possibility of choosing the optimal temperature for both reaction and detection. The response time of the sandwich-type sensors, being diffusion controlled is not very fast, so the flow cell set-up does not necessarily involve longer analysis times. (Fig. 31 shows schematically such a set-up.) As indicated in Table 7,

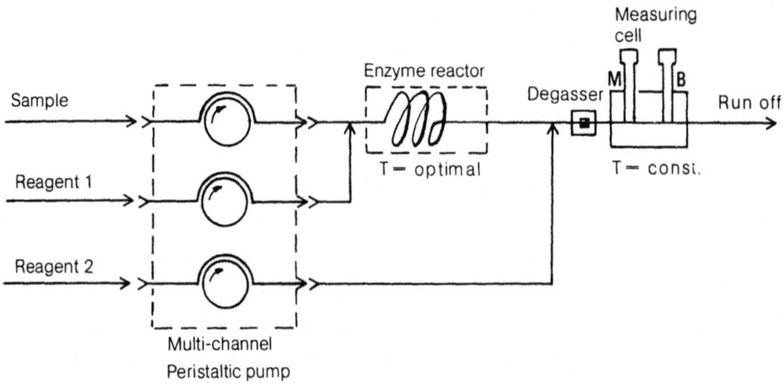

Fig. 31. Flow-thru-bio-sensor. ***Reagent 1*** contains: optimum buffer for the enzymic reaction, stimulating agents, masking reagents for inhibitors, protection agents against bacterial growth. ***Reagent 2*** contains: optimum buffer for the indicator electrode (for quenching, the buffer capacity should be higher than that of reagent 1), if necessary, ions for an indirect monitoring (e.g. iodide + Mo(VI) catalyst for all enzymic reactions which deliver H_2O_2)

a number of biochemical reactions can be carried out in sequence to produce an ion which can be selectively detected. This approach is also no problem for flow cells with reactors aligned in succession.

Concerning the analyzable concentration range, care must be taken not to work with too high a substrate concentration. The nearly Nernstian response is obtained only in concentration regions in which the Michaelis-Menten kinetics are followed. With substrate saturation (reaching the maximum rate), the EMF no longer follows the substrate concentration.

4 Measuring Techniques with Ion-Selective Electrodes

4.1 Equivalent Circuit for a Cell with Liquid Junction

Our task is to measure the EMF of an electrochemical cell as accurately as possible. The measuring apparatus chosen should indicate only the most electrochemically meaningful precision. Fig. 32 illustrates the measuring circuit of an ion-selective in-

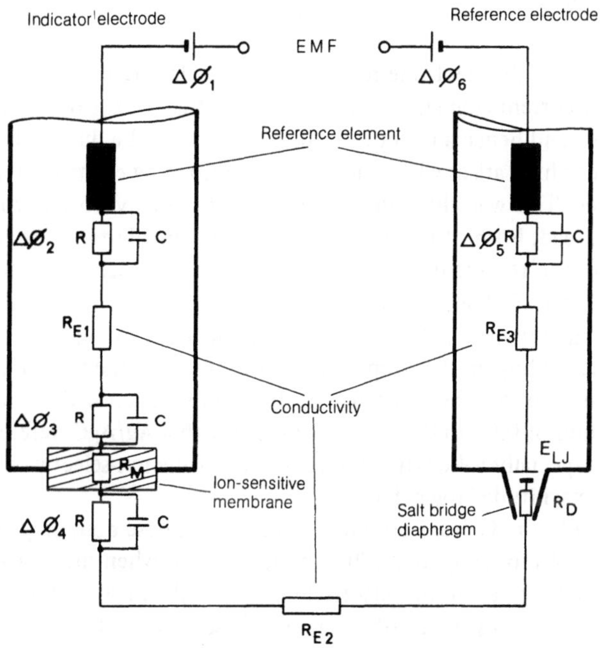

Fig. 32. Equivalent circuit for an electrochemical cell with liquid junction

dicating electrode complemented with a reference electrode in a cell with liquid junction. The potential difference we want to measure is $\Delta\phi_4$. For the reasons mentioned in Chap. 2.1, we must be satisfied with measuring only a change in its value. If a measured change in $\Delta\phi_4$ is to be related to a corresponding change of activity of the measured ion via the Nernst equation, then all the other circuit components must remain constant. To appraise the errors involved in such a measurement, it will be useful to study each of these components in greater detail.

$\Delta\phi_1$ is the Galvani potential at the phase boundary between two different metals, for example at the contact point between the common silver wire shunt element and the copper instrument lead. Ignoring the surface potentials (see $\Delta\chi$ Chap. 1.1), the EMF at such a contact point is known as the contact Volta potential. These can be highly temperature dependent, as demonstrated by the analogous thermocouples. The Ag/Cu contact has a temperature coefficient of 10^{-4} mV/°C, the Cu/tin solder contact 10^{-3} mV/°C, and the Cu/CuO contact an observable coefficient of 1 mV/°C. This last contact may arise if corroded copper plugs are used. In precise measurements striving for accuracies and sensitivities $\leq$ 0.1 mV, clean contacts are a must, and temperature variations should be avoided as much as possible.

$\Delta\phi_2$ is the equilibrium Galvani potential at the internal reference element (usually Ag/AgCl)/inner solution interface. This depends above all on the activity of those ions in the inner solution which take part in the electrode reaction at this element (with Ag/AgCl, the chloride activity). $\Delta\phi_2$ can only be considered a constant if this activity is constant. For the significance of the $R \cdot C$ term at this and all other phase boundaries, refer to Chap. 1.1. R represents the interfacial resistance, which can have many different origins. It may reflect a polarization resistance, which results from a depletion of the ions participating in the electrode reaction near the electrode surface. This depletion, caused by the current (ion-stream) across the interface, leads to a lower conductivity in this region, and hence a larger resistance. R may also be due to the so-called reaction resistance, which arises when the current-maintaining electrode reaction is kinetically hindered. This was illustrated in Chap. 1.3 by the hydrogen evolution reaction at mercury or lead. Both of these phenomena may contribute to the total resistance of this interface. C represents the capacitance of the electrochemical double layer described in the model of Chap. 1.1. This is then connected to the resistance R_{E1}, determined by the conductivity of the inner solution. Since R can take on tremendously different values, depending on the properties of the interface (the exchange current density i_0 is a measure of R), the resistance of the inner solution with the usual electrolyte concentration of 0.01 to 0.1 M (NaCl or KCl with a Ag/AgCl reference) can be neglected. This is especially valid when the resistance of the ion-selective membrane is many orders of magnitude larger than R_{E1}.

$\Delta\phi_3$ is the equilibrium Galvani potential at the inner phase boundary of the ion-selective membrane. As for $\Delta\phi_2$, this will only be constant when the activity of the corresponding ion taking part in this electrode reaction (the measured ion) is constant in the inner solution. The resistance of the active phase, R_M, is often the largest of all resistances in the circuit. As such it determines the error in the EMF measurement due to the $i_m \cdot R_M$ potential drop, along with the circuit current i_m, determined by the input resistance of the voltmeter.

$\Delta\phi_4$ is the *variable to be measured.* It is the change in Galvani potential at the outer phase boundary of the ion-selective membrane. Here R and C depend on the properties of the sample solution and can vary accordingly. The electronic delaying term $R \cdot C$ plays an especially large role in the investigation of solutions with rapidly changing measured ion activity. Such situations arise, for example, during kinetic measurements in physiological investigations, where one is interested in the quickest response possible (see C-neutralization in Chap. 6.1.2.4).

The resistance of the sample solution R_{E_2} lies between the Galvani potential $\Delta\phi_4$ and the equilibrium Galvani potential at the outer reference element $\Delta\phi_5$. In some cases this resistance can be *larger than the resistance of the ion-selective membrane.* This occurs, for example, when working with some pure solvents (no additional electrolyte). In testing the purity of distilled water, for example, irreproducible or drifting values are often obtained. If the resistance of the sample solution exceeds the input resistance of the voltmeter, then it looks to the instrument as if both electrodes were hanging in air. As salt bridge electrolyte leaks out of the reference electrode diaphragm increasing the conductivity of the sample solution, the measured EMF will drift. At first the $R_{E_2} \cdot i_m$ potential drop across the sample solution is very great. With increasing conductivity of the sample solution as the salt bridge electrolyte ions from the reference electrode solution diffuse into the sample solution, this potential drop gradually diminishes until the true EMF of the cell is measured. *In such media, the addition of an inert electrolyte can greatly facilitate the measurements.*

In cells with liquid junction, as illustrated in Fig. 32, the liquid junction potential E_{LJ} and the associated diaphragm resistance R_D (usually < 1 KOhm) also lie between the two equilibrium Galvani potentials $\Delta\phi_4$ and $\Delta\phi_5$. The variations in this liquid junction potential and the methods available for minimizing it have already been discussed in Chap. 2.3. The potential drop $i_m \cdot R_D$ can also be neglected in routine laboratory operations with small circuit currents ($i_m < 10^{-12}$ A), especially since the resistance of the ion-selective membrane is usually significantly larger than R_D. However, if larger stray or leakage currents flow through this diaphragm, but not through the ion-sensitive diaphragm (as is the case in some industrial applications with grounded solutions), then an unrealistically high potential drop may result which will interfere with the measurement. The same situation arises if the diaphragm becomes clogged up, causing the resistance R_D to increase substantially. For these reasons it is advisable for very precise measurements in cells with liquid junction to use a sleeve diaphragm which can easily be kept clean. In poorly conducting media, the high outflow of salt bridge electrolyte through this type of diaphragm quickly provides the necessary sample solution conductivity.

The equilibrium Galvani potential at the interface: Reference electrode element/salt bridge electrolyte, $\Delta\phi_5$, is encountered after the salt bridge solution resistance R_{E_3}. This potential depends, as do all equilibrium Galvani potentials, on the activity of the ion participating in the electrode reaction. Since this activity is held constant by the salt bridge, $\Delta\phi_5$ can also be considered constant. One must pay attention, however, to any polarization resistance arising at this interface (unpolarizable electrode, see Chap. 1.2). As we have seen, especially in industrial applications using grounded solutions, detectable stray currents may flow across this interface, significantly altering its equilibrium Galvani potential. The last component of this equivalent circuit is the Galvani potential $\Delta\phi_6$ arising at the contact point of two different metals. The same considerations which were mentioned in discussing $\Delta\phi_1$ hold here. If the same metals are used, $\Delta\phi_1$ and $\Delta\phi_6$ can completely cancel one another.

This detailed analysis of a measuring circuit is merely an introduction to the complexity of a practical EMF measurement, which on the surface may appear deceptively simple. As long as an accuracy of ±5 mV is sufficient, many of the points discussed here can be ignored. However, it is often the case when working with ion-selective electrodes that one wants to make full use of the accuracy of the electrochemical cell in order to obtain correspondigly precise and reproducible analytical results. *This understanding of the properties of an electrochemical circuit emphasizes the futility of attempting to increase the accuracy of an analysis merely by using a more sensitive and precise voltmeter!*

4.2 Measuring the EMF of an Electrochemical Cell

The last section, along with the discussions of exchange current density and polarizable and unpolarizable electrodes in Chap. 1.3, show how important it is to carry out the EMF measurement between two electrodes with as little current as possible. Connecting the electrode leads directly to a moving coil meter, similar to the widely used multimeters, without electronic amplification allows the measurement of the true EMF only in the rarest cases (very high exchange current densities at both electrodes, very low inner resistance of the electrochemical circuit). In principle, the potential difference between two terminals can always be measured with an appropriate galvanometer, but whether this value represents solely the resting EMF of interest is questionable. The requirements of a potential measurement using the smallest possible current flow can be illustrated by considering the potential measurement of a charged capacitor. Figure 33 illustrates the processes which occur in attempting to measure the potential across a charged capacitor using a low-resistance voltmeter. Such a measurement is actually a

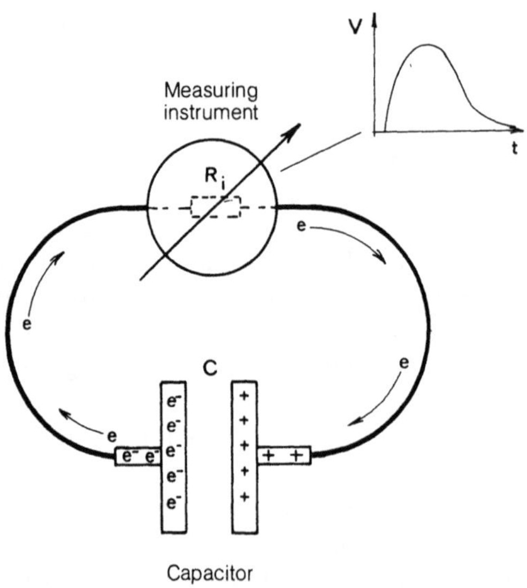

Fig. 33. Voltage measurement across a capacitor

way of short-circuiting the two capacitor plates. The measurement immediately leads to an equilization of the charges on both plates. Only a transient signal (current impulse) is observed until the capacitor has discharged through the measuring coil. The work which must be done to measure the potential across the charged capacitor plates in this case is too large relative to the work stored in the system. It is this stored work which we are actually trying to measure via the potential determination. Complete elimination of any current flow during a potential measurement is hardly possible on the other hand, since the detecting instrument interacts with the system only through a current flow. The test circuit current and the current through the membrane need only be small in comparison to the smallest exchange current density i_0 at the electrode surfaces, so as not to disturb the energy equality:

$$\text{chemical energy} = \text{electrical energy}.$$

In contrast to ion-selective electrodes, electrodes of the first and second kind, as well as redox electrodes, usually exhibit exchange current densities $i_0 > 10^{-3}$ A/cm^2. Such electrodes can easily tolerate currents of around 10^{-6} A. For this reason, not nearly as much importance has been attributed to the circuit current with earlier electrochemical techniques. Many textbooks mention the compensation method of Poggendorf in this connection. In this method the EMF to be measured is balanced by an equally large opposing voltage to avoid too large a circuit current. Figure 34 illustrates this

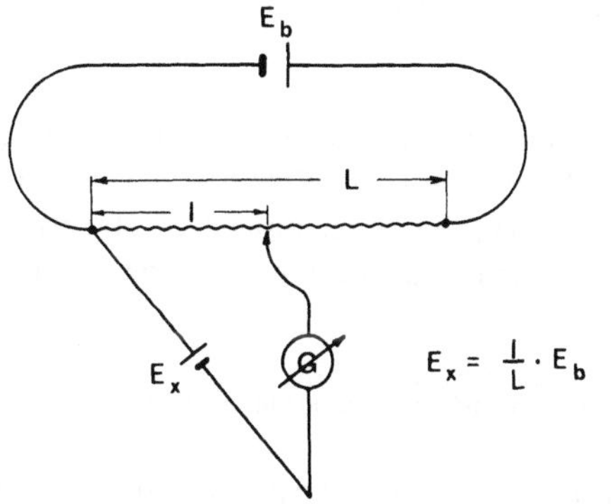

Fig. 34. EMF measurement according to the compensation method of Poggendorf

principle. The compensating resistance of the slide-wire potentiometer is varied until the sensitive galvanometer G indicates no further current flow. At this point both voltages are equal, and by using a standard cell (for example Weston = 1.01841 V), the electrode EMF can be calculated directly from the measured ratio of the slide-wire distances l and L. In this method the galvanometer need never be fully deflected, thus lowering the current needed to make the measurement. However, this method is not suitable with most ion-selective electrodes for two reasons. First, the method does not guarentee extremely small measuring circuit currents ($i_m < 10^{-12}$ A) and second, when the measuring circuit is not exactly balanced, it is short-circuited through a portion of the slide-wire resistance. The resulting current will induce non-equilibrium conditions at the electrode surface and can lead to drifting results.

4.3 Choosing an EMF Measuring Device

For the sake of clarity, the detailed equivalent circuit shown in Fig. 32 can be more simply represented. In Fig. 35, R_S represents the sum of all test circuit resistances and is designated source resistance.

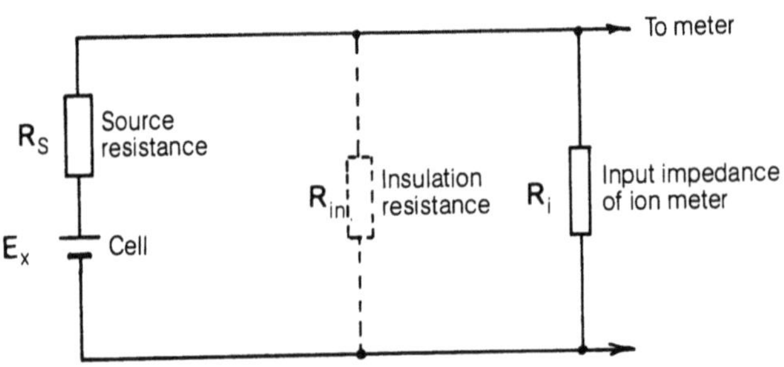

Fig. 35. Electric circuit diagram of the amplifier input

R_i represents the input impedance of the ion meter used. According to this simplified circuit, the electrode EMF is short-circuited by the input impedance if this impedance is substantially smaller than the source resistance. The error resulting from poorly matched resistances can be calculated with the following formula:

$$\%\ \text{Error} = \frac{\text{source resistance}}{\text{source resistance} + \text{input impedance}} \times 100. \tag{48}$$

In order to keep this error below 0.1%, the input impedance of the indicating instrument must be larger than the source resistance by a factor of 1,000. With ion-selective electrodes resistances of up to 1,000 MOhm must be accommodated (glass membrane electrodes at temperatures $< 10°C$). This means that the input impedance of the instrument must be $> 10^{12}$ Ohm. Among potential measuring devices, such a high input impedance is exhibited only by the so-called electrometers. Special "pH meters" designed specifically for measuring the EMF of cells containing pH glass electrodes have been available since 1935. In addition to a mV scale these also display a pH scale, which is pre-calibrated by the manufacturer according to the temperature-dependent Nernst equation (at 25°C, 59.16 mV per pH unit). Naturally, this scale can also be used to indicate the corresponding pIon values of other monovalent cations when used with an ion-selective electrode. With divalent cations the Nernst slope is only half as large (at 25°C, 29.6 mV/power of ten).

If anions were measured, the indicator would be deflected in the opposite direction. A few newer instruments have been designed specifically for use with ion-selective electrodes. These have controls which enable the operator to specify the type (anion or cation) and valency (mono- or divalent) of the indicated ion, so that the pH scale can be used directly without further calibration. Often a logarithmic scale is provided as well to allow direct indication of activities and concentrations (Fig. 36). An ion meter can

Fig. 36. Ion meter with analog scale and circuit board programming (Polymetron)

be divided into three sections. The first is the input circuit, whose main function is that of an impedance transformer. Currently available instruments contain predominantly field-effect transistor or capacitance diode inputs. The second section consists of the power amplifiers. The power requirement of the indicating device is usually much higher than the amplified power output of the first section. Since this section has already effected an impedance transformation (high-ohmic → low-ohmic), the power amplifier can be constructed using low-ohmic circuit techniques. Depending on whether the EMF to be measured has been modulated before reaching the input circuit (AC amplifier) or not (DC amplifier), an electronic rectifier circuit to correct the phase of the AC signal can also be included in this second section. The third section consists of the indicating meter, as well as any other functional devices (on-off, mV-pH, calibration knob, temperature compensator, % Nernstian behavior). Further distinctions can be made according to the type of display used: Analog-indicating instruments with and without scale expansion, digital display instruments and compensating instruments, which can indicate the position of the potentiometer (compare with the Poggendorf circuit) in either analog or digital fashion, depending on the scale.

Which of these possibilities one should choose depends upon the particular type of application. The analog-indicating amplifiers are the least expensive. This type of indicator employs a moving coil meter. In order to make full use of the visual reading precision of the scale (~ 10 to 20 cm), a meter with ribbon suspension should be used; otherwise errors due to non-linearity may be greater than 0.5%. The visual precision can be doubled with the use of a "folded" scale. Such meters have two indicating ranges (0 to 8 and 6 to 14 pH units) with a switch to determine the working range. This type of meter is especially suited for battery operation, especially when used with integrated circuits and operational amplifiers. The visual precision of such meters usually lies between ±0.03 to ±0.05 pH units, or ±3 to ±5 mV. This corresponds to an error in activity of about 15 to 20% for monovalent ions, and about twice as much for divalent ions. For routine analyses and monitoring purposes (such as geological trend analyses in the field) this is usually sufficient.

For greater precision, a better instrument must be used. Meters with analog display, but which have additional expanded scales belong to the next higher price range. With such devices the full scale can be used to represent 1.4 or 2.0 pH units. The zero point of the expanded pH or mV scale can be dialed in with a calibrated step switch on some instruments. Such scale expansions are operationally partial compensation processes. A portion of the amplified output signal is compensated for by a constant opposing voltage whose magnitude is determined by the precise, linear resistances in the voltage divider operated by the step switch. Only the small remaining portion of the original signal is indicated. The error in linearity of the indicating instrument no longer enters into the full measured result with this design, but rather only into the small displayed portion. A prerequisite for improved accuracy, of course, is that the compensating potential be correspondingly accurate. pH meter amplifiers with expanded scales allow a visual precision of ±0.01 pH unit, or ±1 mV. This precision is quite sufficient for routine analyses using cells with liquid junction, since the electrochemical cells themselves do not provide a substantially greater accuracy in most cases (see Chap. 2.3 concerning the liquid junction potential). Although the scale is formally increased by a factor of 7 to 10 with this design, the calibration and sample solutions should not differ by many pH units for purely electrochemical reasons. Instruments with scale expansions, but without the precisely calibrated zero suppressing step switch can be purchased at only slightly higher prices than those with "folded" scales.

For yet higher precisions, which can only be fully utilized by cells without liquid junction, one must select a compensating or digital apparatus. The compensating meters operate on the principle of the Poggendorf compensation circuit. Here, however, the signal is compensated after impedance transformation and amplification. Since potentiometers can be linearized to better than 0.05%, these devices yield a high accuracy over the entire range. With potentiometers having scales which can be divided into many small regions, total scale lengths of many yards result, guaranteeing a visual precision of ±0.001 pH unit, or ±0.1 mV. A disadvantage of the compensating meter is the longer time required to complete a measurement due to manual zeroing of the potentiometer (about 10 to 60 seconds, depending on the skill of the operator). Such zeroing is difficult, if not impossible, with drifting EMF values. For this reason these meters are less often employed for routine measurements. Instruments are also available which automatically zero the potentiometer via a servo motor (E.I.L.).

Digital pH meters are equally well-suited for routine and precise measurements. Their greater accuracy (about 0.1%) and clear readout make these the instruments of the future, especially since they are already available for less than $1,000 (see Fig. 37).

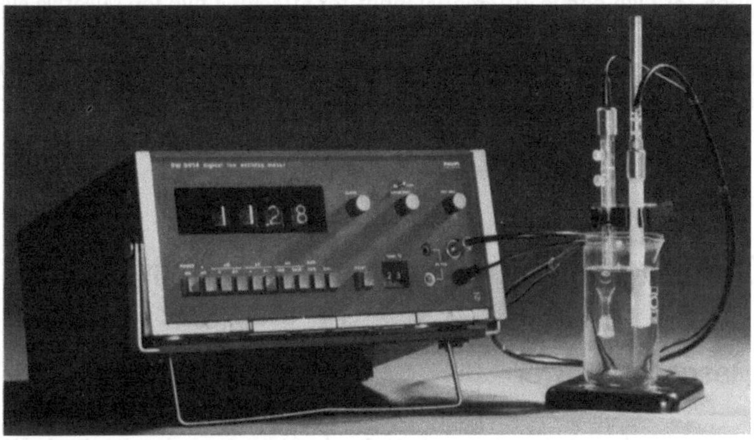

Fig. 37. Digital pH/pIon meter (Philips)

Determining the desired final potential value, indicated only after equilibrium has been established and any drifting ceases, is equally difficult with all indicating methods. *To get the full benefit of the precision provided by any measuring instrument, the appropriate time period must elapse before recording the data.* A chart recorder connected to the potentiometer is often a great help in determining when this point has been reached. Such a trace of the EMF development in many cases makes it far easier to decide whether or not a "near final" potential value has been reached. A small, inexpensive recorder is fully sufficient for such applications.

Finally, a few more requirements for a practical ion meter are enumerated, along with a summary of those already discussed [213]:

1) Ease of Operation
   Since often people who don't even know the charge of the measured ion are entrusted with the job of making measurements, a circuit board with the proper programming should be supplied with the instrument (see Fig. 36). From the standpoint of the customer service it might also be suggested that such circuit boards also adjust for the corresponding isotherm intersection point. In this case the entire flow system must be purchased from the same manufacturer.
2) Direct concentration indication in digital form for direct potentiometric measurements
3) Concentration indication by the method of standard addition (for example, increment scales on analog instruments)
4) Input impedance $> 10^{12}\ \Omega$
5) Error of $\leq \pm 0.1$ mV over a total working range of ± 1.5 V
6) Concentration indication to three significant figures
7) Temperature setting reproducibility of 0.2°C
8) Isotherm intersection point adjustment for industrial instruments ($E_{iso} \pm 10$ pActivity units)
9) Zero suppression control with ±1,000 mV range
10) Drift $<$ 1 mV/24hour

11) Option for high-ohmic differential input for attaching two ion-selective electrodes (see Fig. 43)
12) Option for impedance transformation very near the high-ohmic electrode with input impedance $> 10^{14}\ \Omega$, guarded signal and capacitance neutralization for physiological measurements.

4.4 Characteristics of Electrometer Amplifiers

4.4.1 Resolving Power

Every amplifier can be rated according to its level of background noise, i.e. indicator fluctuations (usually noticeable only in the most sensitive range). This is the sum of the noise levels of each of the individual components (transistors, tubes, resistances, etc.). Its origin lies in the statistical irregularity of the electron motion (current). The electronic amplification of a measured signal is limited, because at too high amplifications the signal disappears in the background noise of the amplifier used. Usually measured signals themselves have some noise level. The signal to noise ratio of the signal from the EMF source cannot be further improved through electronic amplification. The electronic amplification of a measured signal is limited by the point at which the source noise causes indicator fluctuations. As previously discussed, EMF measurements with ion-selective electrodes involve high source resistances. In an electric circuit, however, *each resistance functions as a noise generator.* The higher the resistance, the greater the noise. The molecules of the resistor material transmit their thermal energy (oscillations about their equilibrium lattice positions) to charged particles (ions and electrons) through collisions, leading to statistically distributed elementary current impulses (current = movement of electric charges). This is called "thermal resistance noise" of a resistor. For a metallic conductor this can be estimated with the following formula [214]:

$$E_r = \sqrt{(h \cdot f/k \cdot T)\,[\exp(h \cdot f/k \cdot T) - 1]^{-1} \cdot 4k \cdot T \cdot R \cdot b} \qquad (49)$$

or at $T = 300$ K, approximately:

$$E_{r,\text{effective}} = 0.13\ \sqrt{R \cdot b}\ [\mu\text{V} \cdot \text{k}\Omega, \text{kHz}]. \qquad (50)$$

Where:

h = Planck's constant	T = temperature
f = frequency	R = resistance
k = Boltzmann constant	b = bandwidth

Figure 38 shows the thermally predicted (according to Eq. 50) peak to peak voltage noise level in [V] as a function of the source resistance R and the bandwidth b of the

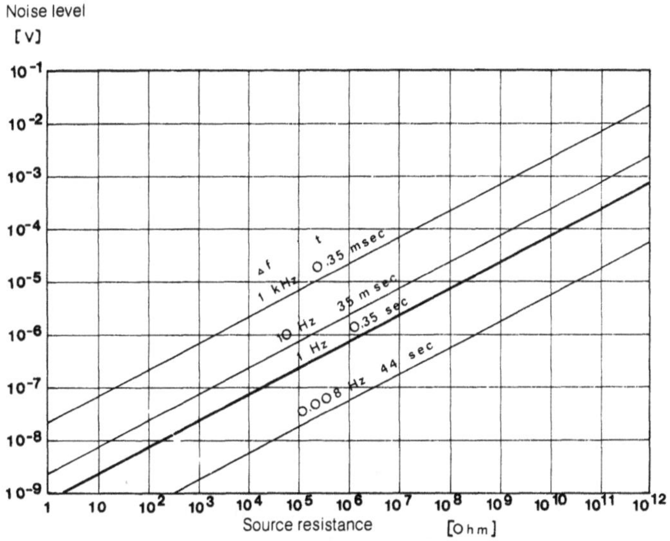

Fig. 38. Voltage noise of resistors as a function of the amplifier bandwidth [214]

amplifier. The bandwidth is the frequency range which is completely amplified. For reading the amplified signal off of a moving coil meter, this can be assumed to be about 1 Hz. This graph shows that for a metallic resistance between 10^6 and 10^9 Ohm, the theoretically expected optimum resolution is no better than 0.001 to 0.01 mV. With ion-selective materials one must deal with resistances many times these values. According to Eq. (50), the noise can be decreased by using amplifiers with smaller bandwidths, b. The bandwidth of the amplifier can be lowered by connecting a damping R-C low-pass filter in series, but this leads to impractically long waiting periods, since the time constant of the amplifier is increased by the same amount. (A bandwidth of 1 Hz corresponds to a response speed of about 0.35 seconds to reach 90% of the final value).

In addition to the source noise, the amplifier noise itself must be considered. With the best electrometer amplifiers with short-circuited input and in the most sensitive range a noise level corresponding to ±25 μV can be observed on the indicating meter. This means that the meaningful resolving power of currently available electrometers is about 0.02 mV. [When reading claims that some reference electrodes agree with one another to within better than 0.01 mV, the reader must realize that the source resistance of a cell containing two reference electrodes is considerably lower than that of a cell containing an ion-selective electrode. Therefore less high-ohmic amplifiers (compensators) can be used for EMF measurements in this case].

4.4.2 Insulation Problems

In the simplified equivalent circuit of Fig. 35, another resistance R_{in} can be included in parallel with the electrometer input to represent the insulation resistance of the shielded instrument leads (resistance between the cable core and the shielding, see Fig. 40). In order that the EMF to be measured not be short circuited through a detour to a ground connection (the reference electrode is usually connected to the circuit ground), this resistance must be many orders of magnitude greater than the source resistance. This requirement is exactly analogous to that of the input impedance described by Eq. (48). This means that for a source resistance of 10^{10} Ohm, the insulation resistance should be $> 10^{13}$ Ohm. *To maintain such a high resistance over the entire cable length including the plug connections requires considerable care. Even small amounts of electrolyte (such as CO_2 from the air) dissolved in condensed atmospheric moisture can give rise to interferences during high-ohmic measurements.* One method of avoiding this type of short-circuiting is to keep the measuring apparatus always switched on. The heat generated by the instrument in this case inhibits extensive condensation on the input plugs or in the high-ohmic circuit board electronics. When using high-ohmic ion-selective electrodes with cables having plug connections, the plug insulation should be cleaned from time to time with water-free methanol (reagent grade). An insulation interference is difficult to detect, because the indicator of the meter does not fluctuate as it does with other interferences. This type of interference merely results in a smaller measured EMF. pH simulators (such as the Model E448 of Metrohm with a calibration accuracy of 0.1% in either mV or pH units, with a source resistance of 10^9 Ohm) are suitable for testing the proper function of such an instrumental set-up. The electrode cable resistance can be checked with an electrometer multimeter having an Ohm-measuring range up to 10^{14} Ohm (Keithley, Princeton Applied Research).

4.4.3 Charging Phenomena

Almost all users of pH meters are familiar with static charging phenomena. These can be explained with a modified form of the simplified equivalent circuit of our electrochemical cell (Fig. 39). Here an electrode cable capacitance C is introduced in parallel

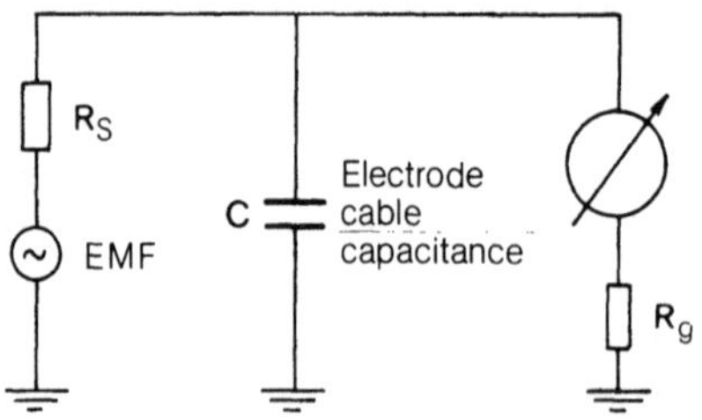

Fig. 39. Equivalent circuit of the amplifier input taking into account the electrode cable capacitance

with the electrometer input. As an indication of its magnitude, the capacitance of an electrode surface is between 0.1 and 10^3 μF/cm^2; that of one meter of shielded electrode cable is about 0.001 μF. Before the charging of the electrodes reaches the electrometer input, it must charge the capacitor formed by the cable core and the associated shielding. This in turn creates a voltage U across the capacitor plates (cable core and shielding) according to

$$U = \frac{q}{C}. \tag{51}$$

U is the same as the EMF of the electrochemical cell. The capacitance C depends on the geometry, i.e. the spacial arrangement of the two condensor plates (cable core and shielding). The magnitude of the charge q which builds up on the capacitor plates according to Eq. (51) can only be changed very slowly with extremely high-ohmic electrometer circuits, because only the very minutest currents are flowing in the test circuit. If the geometry of the electrode "cable capacitor" is suddenly changed (by bending, for example), the capacitance C is altered, as well as the charge required to establish the voltage U according to Eq. (51). Since in the case of extremely high-ohmic test circuits this charge can only very slowly establish its new value, it temporarily determines the voltage registered by the electrometer. Only after a definite period of time is the charge equilibrium re-established through the extremely small test circuit current and the cell EMF the same as the voltage across the "cable capacitor". When working with high-ohmic electrodes (glass membrane electrodes above all), this results in temporary meter deflections whenever the electrode cable is disturbed. With low-ohmic electrodes, such as redox electrodes or electrodes of the first kind, this effect is not so pronounced, since the charge equilibration can occur so fast through the low-ohmic electrode circuit that nothing is noticed on the electrometer indicator. For this reason such electrodes rarely have shielded cables. In addition to the observable effects of changes in the electrode cable capacitance, high-ohmic circuits are also very sensitive to electrostatic induction. Any charge, upon approaching a material phase, induces an equal and opposite charge in this phase. In this way an *external* charge may influence the test circuit and evoke a variation in the circuit charges. Since this additional charge cannot very quickly dissipate in high-ohmic circuits, it leads to a voltage change between the cable core and the outer reference electrode. This voltage is indicated by the electrometer until the charge is dissipated throughout the test circuit. *Especially troublesome in this connection is an electrostatic charging of the operator (enhanced by insulating shoe materials and synthetic fiber clothes).* Considering that such charging can sometimes even lead to spark formation ($>$ 2,000 V), it is easy to understand its influence upon extremely high-ohmic EMF measurements. The signal-carrying cables are shielded in an attempt to block out the effects of such strong external charges. Thus the signal conductor is surrounded with a conducting sheath connected to the electrometer circuit ground. In this way charging is induced primarily in this shielding, and having a small resistance, this conductor can quickly dissipate the charge to the circuit ground. At any points where this shielding is interrupted, electrostatic induction is still a problem. *For this reason ion-selective electrodes having high-ohmic active phases should have shielding which reaches as closely as possible to the electrode tip.* Although an electrolyte solution (sample solution) often displays shielding behav-

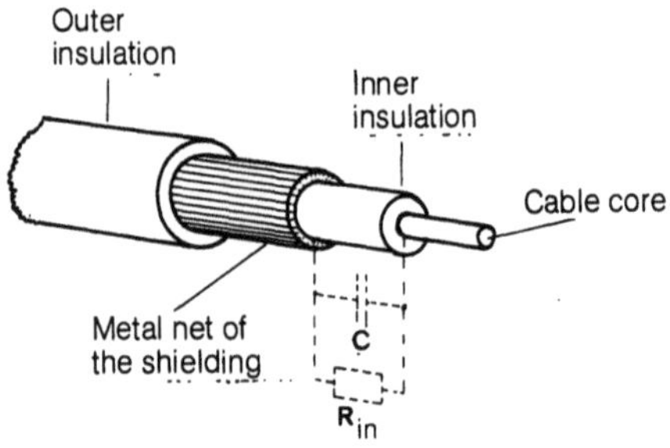

Fig. 40. Construction of a shielded cable

ior, difficulties may still arise in this connection when using diluted, poorly conducting solutions. With some of the commercially available liquid ion-exchanger electrodes, stable electrometer readings can only be obtained by surrounding the electrode body with aluminum foil connected to the circuit ground. If even this does not help, then the entire apparatus must be placed in a suitable Faraday cage.

As a rule, the low-ohmic *reference electrode* is connected to the low-ohmic side of the electrometer input, so that any charges induced in this branch of the circuit can be immediately dissipated. For this reason the reference electrode cable is not shielded. Exceptions are reference electrodes which are used in conjunction with potentiostatic set-ups, since in this case they are connected to the high-ohmic side of the electrometer amplifier. With a low-ohmic *indicating electrode* (< 1 MOhm) electrostatic induction does not interfere very much, since the charge can flow unhindered to the circuit ground through the electrochemical circuit (through the electrolyte solution and the reference electrode).

One other possible interference remains: The *cable noise.* This effect can arise if movement of the shielded electrode cable causes the shielding network to rub against the polyethylene or teflon insulation, producing frictional charges. The so-called "low noise" cables can help to alleviate this problem. These contain a polyethylene insulation (to avoid the piezoelectric effect of teflon) which is coated with a layer of graphite powder or conducting synthetic resin.

When working with high-ohmic indicator electrodes the cable connections should be stable and free of vibrations. In addition, no current-carrying power cords should be positioned very close to the electrode cables. The shielding used is not very good for magnetic fields, so the fields associated with the power cord can induce alternating signals of the power source frequency in the test circuit. Such induced signals may not be noticed due to the sluggishness of the indicating instrument. Nonetheless, these signals can easily bring the amplifier into a saturated region, in which it no longer exhibits linear amplification characteristics. When using an oscilloscope to follow the EMF (as in some physiological applications), this power frequency hum becomes visible and extremely interfering.

4.4.4 Ground Loops

Ground loops can be formed if a number of instruments with individual power connections are galvanically connected together. When working with ion-selective electrodes, this usually means a recorder or oscilloscope is connected to the recorder output of the electrometer amplifier. Ground loops are usually overlooked. Often these interferences are blamed on the apparatus, which is "having a bad day", or on the "poor wall socket ground", since a different socket previously yielded better results, etc. Figure 41 illus-

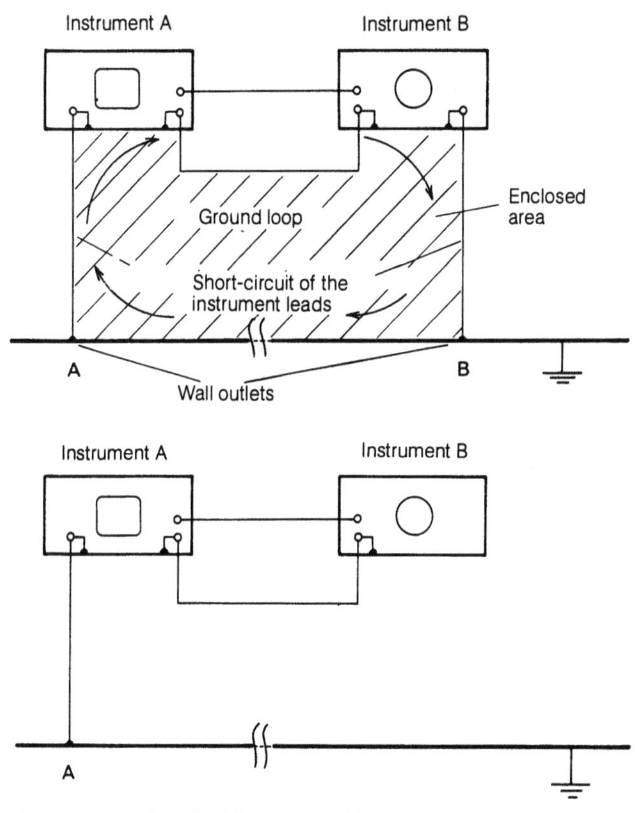

Fig. 41. Measuring set-ups with and without ground loops

trates what is meant by the term ground loop. In this figure, only the ground connections are shown. It is easy to see from this simplified picture that the two galvanically connected instruments are part of a large closed circuit (the ground loop) completed by the two power cords and the connection between the two wall sockets. The galvanic coupling of the two instruments can arise in many ways. In some testing set-ups the instrument housings are in direct contact; a shielding conductor connected to the ground has the same effect. Depending on the circuit, the loop can also be closed through the circuit ground (often labeled the minus output or input plug; in some instruments the circuit ground and the housing or shielding ground are separated by a 1 Ω resistance). A current will flow in this circuit if either 1) a galvanic activity (such as a contact potential) is present in the circuit, 2) a voltage difference exists between the circuit grounds of the connected instruments, or 3) most disturbing due to its inconsistency and hum, alternating electric or magnetic fields which penetrate this loop (principle of the dynamo). Due to the often considerable size of the ground loop (with some wall socket connections this may encompass several rooms) the alternating fields of the house line frequency present almost everywhere are the most strongly interfering. The input amplifiers of most instruments made today are built with specific circuits to suppress this frequency. This common mode rejection is limited to a small frequency range (about 50 Hz) however, so that alternating fields of different frequencies (from electrical switches, unscreened motors, etc.) can still lead to unstable readings with large ground loops. *The only solution to this is breaking the galvanic contact or reducing the size of the ground loop.* Thus all instruments which are connected and conduct the measured signal should be supplied by the same power outlet, and the area spanned by the power cord should be kept to a minimum. Especially useful in this context is the multiple outlet E 550 of Metrohm, which in addition contains an interference suppressor to reduce interfering signals in the frequency range 1 KHz to 1 MHz.

5 Analysis Techniques Using Ion-Selective Electrodes

This section describes the various calibration and evaluation methods which can be used with ion-selective electrodes. *The specific procedure chosen determines whether the activity or concentration of the indicated ion is obtained as a result of the measurement.* Fundamental to all analytical methods is the Nernst equation. We have seen in Chap. 1.2 that the Nernst equation is a theoretically derived relationship. *In practice it is usually the case that an electrode deviates somewhat from the Nernst equation. As long as this deviation is reproducible, the electrode can still be used for analytical applications.* Testing an electrochemical cell for exact Nernstian behavior is quite painstaking, and is often impossible. Standard solutions of precisely known activities are needed. This, in turn, requires the individual activity coefficients of the corresponding measured ions, and these are not always known with sufficient accuracy. Instead, one usually uses an empirical form of the Nernst equation for analytical applications:

$$E = E' \pm S \cdot \log a_M \qquad (52)$$

where: E = EMF of the electrochemical cell

E' = EMF of the electrochemical cell with a solution in which $\log a_M = 0$; i.e. $a_M = 1$

$\pm S$ = slope, (+) with cations, (–) with anions (at 25°C, theoretically = 59.16 mV/pActivitiy unit for monovalent ions)

Equation (52) can be drawn as a straight line using a logarithmic activity axis. A line is uniquely described by two specifications: either two points, or one point and the slope. *Thus each ion-selective electrode should be calibrated with at least two standard solutions if the activity or concentration is to be obtained directly from the EMF measurement (direct potentiometry).* For titration endpoint determinations an exact calibration is not required, since the desired value is only the point at which the greatest change in EMF occurs. Which analytical technique is used depends on a number of factors. If simplicity and speed are crucial, then a direct indicating procedure is employed. If more value is placed on accuracy, then a titration procedure is more appropriate. In addition, there are combinations of both procedures, such as the addition method (internal standard) and the method of Gran. It must be noted here, that in all of the following sections it is assumed that the standard solutions and sample solutions are at the same temperature (thermostated).

5.1 Calibration Curves

Drawing a calibration curve is by far the most easily overlooked procedure for ascertaining the activity or concentration of an indicated ion in a sample. Even if another evaluation method is being employed, the function of the electrode should periodically be tested by means of a calibration curve. As a rule, one starts with a 0.1 M solution of the indicated ion (0.01 M in the case of the cyanide electrode, to minimize attack on the active phase) and progressively dilutes it until no further dilution effect can be observed by the detecting instrument. The calibration curve for the respective electrochemical cell is obtained by plotting these measured EMF values on a linear ordinate scale against the concentrations on a logarithmic abscissa scale. The *lower detection limit* is taken to be the measured ion concentration at which the measured EMF values begin to deviate from the calibration line. The lower detection limit with ion-sensitive detection techniques is defined analogously to that for other physical chemical techniques, as the measured ion concentration at which the measured signal is exactly twice as large as the background noise. This is the case when the deviation from the Nernst equation is 18/z mV (at 25°C, $59.1/z \cdot \log 2 = 18/z$). All calibration curves derived from concentration values and produced by successive dilutions with water show deviations from linearity at concentrations above about 0.01 M (see Fig. 23). This deviation stems from the fact that the electrode responds to the activity of the ion rather than the concentration. With increasingly more concentrated solutions the activity coefficients take on values progressively smaller than 1.00. In more concentrated solutions of pure measured ion the activity is progressively smaller than the concentration, leading to a leveling out of the concentration calibration curve. If this decrease in activity coefficients is corrected for by using data available in tabular compilations [216], then linear calibration curves are obtained up to concentrations of > 1 M. Before preparing the standard solutions, one should have already decided if one is more interested in the activity (the determining factor in physico-chemical processes such as kinetics, equilibria, etc.; thus more a criterion for the presence or absence of a specific effect) or concentration (of main importance in purity criteria, monitoring toxic substance levels; thus for specifying absolute amounts of material) of the corresponding measured ion.

5.1.1 Determination of Activity Using an Activity Calibration Curve

In the determination of the activity of the free measured ion in a sample matrix, the matrix plays no role in the measurement, provided that it does not contain any other ions which are indicated by the electrode. The only possible effect of the background composition of the sample solution is to influence the liquid junction potential (see Fig. 32) present whenever cells with liquid junction are used. Even the presence of high concentrations of strong electrolytes, usually resulting in small activity coefficients, does not interfere. Complexing and precipitating agents are also of no concern, since only the activity of the free ion is of interest. In such applications a single activity calibration curve can be used without hesitation for all possible sample solutions.

This is in contrast to concentration calibration curves, for which standard and sample solutions should differ in composition as little as possible. This completely unproblematic use of a single calibration curve is counterbalanced by the difficulty of precisely determining the individual activity coefficients of a particular type of ion in the standard solutions. Here again, one starts with concentrations, available by accurately weighing a certain amount of some salt of the corresponding measured ion. *All methods for calculating the individual activity coefficients are only approximations. Thus one must use sufficiently accurate values experimentally determined by an independent method, and these are available only for a few ions* (see Appendix). When this same problem arose in very precise pH measurements, the value of the activity coefficient of a dilute chloride solution was internationally agreed upon [54]. This lead to the operational definition of pH and to the well-defined pH values of a few standard buffer materials. It is to be hoped that similar agreements will be reached in the near future concerning ions currently detectable with ion-selective electrodes [68]. If there is simply no reference concerning the magnitude of the activity coefficients of a particular measured ion, the calibration line obtained with solutions less than 10^{-3} M can be extrapolated to higher values. Some activity coefficients for solutions above 0.1 M have been determined in just this way [58]. In many instances the most recent values for the activity coefficients are included with operating instructions of the particular ion-selective electrode, making the corresponding activity calibration curve easy to construct. In appraising the analytical accuracy of results obtained in this way, one must keep in mind that not only the electrochemical and instrumental accuracies and the standard deviations of the individual measurements must be considered. *Analytical measurements can never be obtained more accurately than the calibration standards are known or defined!*

The actual measurement of the activity of an unknown solution is then quite simple. The EMF of the electrochemical cell containing this solution is measured, and the resulting value is found on the calibration curve. The activity of this measured ion in solution can then be read directly off the abscissa. Normally the stability of an electrode apparatus is good enough that a single calibration curve per day suffices. If higher demands are placed on the accuracy ($< 1\%$), then more frequent calibration checks are required. The slope is usually found to remain constant. In such cases it is sufficient to determine only the parallel displacement of the calibration line, and this can be accomplished with a single standard solution measurement. Experience is the best judge of the conditions under which a new calibration is required. *Care is required especially when working with standard solutions less than 10^{-5} M, since highly diluted, unbuffered solutions are unstable due to irreversible adsorption processes on the container walls. For this reason such solutions should always be freshly prepared before use.*

5.1.2 Determination of Concentration Using a Concentration Calibration Curve

In using concentration calibration curves, it is always assumed that the activity coefficient of the measured ion is the same in the standard and sample solutions. However, this is only the case if the total ionic strengths of the two solutions are the same. There are two different ways to produce solutions as similar as possible. In the first case, the calibration solutions can be prepared with a matrix comparable to that of the sample solution. If the total ionic strength of the sample solution is less than 10^{-3} M, then the sample and standard solutions may differ in ionic strength by as much as a factor of 5 and still maintain an accuracy of ±2% for monovalent and ±5% for divalent ions. This same accuracy requires ionic strengths agreeing within ±50% for solutions between 10^{-3} and 10^{-2} M, and within ±20% for solutions with ionic strengths between 10^{-2} and 10^{-1} M [217]. Such adjustment of the standard solutions to match the sample solution composition can only be realized in cases where the composition of the sample solution is known to some extent, and when no great variations in composition are expected among the various sample solutions. In all other cases a different method of accounting for the ionic strength must be chosen. This involves the addition of enough inert electrolyte to both sample and standard solutions to produce so high an ionic strength, that its variations among the various sample solutions can be ignored. Although it is seldom the case, if the sample solutions have very large ion strength variations, then they are diluted with a concentrated solution of this inert electrolyte (ionic strength adjustor). In this way variations in total ionic strength between 0.1 M and 1 M can be reduced to between 4.51 M and 4.60 M by a 1:10 dilution with a 5 M solution. It has already been pointed out in the discussion of the individual electrodes that this process can be combined with the addition of a suitable buffer system and complexing agents to suppress interferences. Some manufacturers offer such multi-purpose conditioning solutions (e.g. ADE, CONDISOL, TISAB, etc.). Naturally the standard solutions are handled in the same way, and diluted in the same ratio with the conditioning solution. In this way linear concentration calibration curves can be obtained even at high measured ion concentrations, since the activity coefficients remain constant. In addition to allowing the direct indication of concentrations, this uniform composition of the sample solutions also serves to produce a more constant liquid junction potential in cases where cells with liquid junction must be used. A 5 M $NaNO_3$ solution is often used as an inert electrolyte (unless, of course, Na^+ or NO_3^- ions interfere). With sodium-selective glass membrane electrodes a solution consisting of 1 M NH_4OH + 1 M NH_4Cl is the best choice, since it also does the job of establishing the optimum pH. With ClO_4^-, BF_4^- and NO_3^- electrodes, where NO_3^- interferes or is measured directly, a 1 M Na_2SO_4 solution is used as an ionic strength regulator. With the cyanide and ammonia electrodes a 10 M NaOH solution serves well.

This method, use of a concentration calibration curve obtained by dilution with a suitable conditioning solution, spans a broad range (usually 4 to 6 orders of magnitude in concentration) with the same relative accuracy. It is easy to carry out, and in the absence of interferences of the first kind allows analyses with an accuracy of ±2% for monovalent and ±4% for divalent indicated ions.

5.2 Direct Indication on the pH or pIon Scale of an Instrument

In the measurement of monovalent cations, the analogous pIon values (pIon = $-\log a_{\mathrm{Ion}}$) can be read directly off of the pH scale of any pH meter (provided that the built-in offset voltage range is sufficient). With divalent cations or mono- or divalent anions this can only be done with special ion meters (Fig. 42), which allow the sign and valency of the

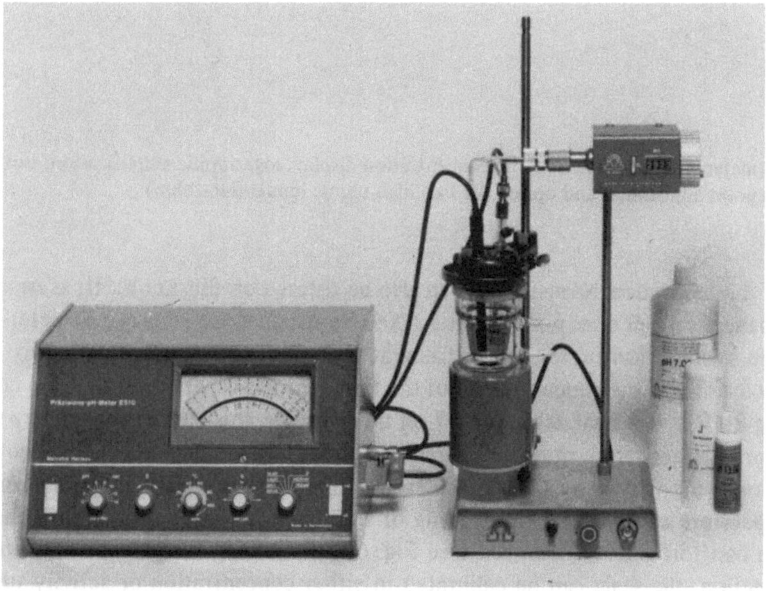

Fig. 42. Ion meter with micro burette (Metrohm)

ion to be chosen with a function switch. There are two possibilities for the scale calibration: Either the pIon values are read off of a linearly divided scale, or the activity of the measured ion is read off of a logarithmically graduated scale in ppm, milliequivalents, etc. (after appropriate calibration with the addition of a conditioning solution, the concentration can also be indicated directly). The last sort of direct readout is this simple only with analog-indicating instruments. Instruments with digital display must use a logarithmic amplifier for direct activity or concentration readout (Fig. 43).

At the basis of every direct indication is the assumption that the instrument scale coincides with the calibration curve. The temperature knob of the pH or ion meter determines the slope of the line, and the calibration or asymmetry control fixes the y-axis intercept (mV value) at a zero value on the x-axis (identical activities of the inner and outer solutions). When the cell EMF is zero, the value indicated on the instrument scale is unaffected by varying the temperature control knob (usually the

Fig. 43. Ion meter with digital activity or concentration display, logarithmic amplification, isotherm intersection point adjustment and option for two high ohmic inputs (Metrohm)

percent of the theoretical Nernst slope can also be entered on this knob). If, as usual, the electrochemical cell does not exhibit a 100% Nernstian slope (this can be determined from a calibration curve), the instrument scale must be calibrated with two standard solutions. If the concentration of the first solution is chosen so that the corresponding EMF is very near zero (pH = 7 on the pH scale; this concentration may be known approximately from previous calibration curves), then the value of the second solution can be brought into exact coincidence with the meter scale by adjusting the slope/temperature control, and the reading of the first solution (adjusted with the calibration control) is not influenced (see Fig. 60). Following the guidelines in the previous section, the scale can be calibrated in either concentration or activity units in this way. The only requirement here is that the calibration curve be linear. Concerning linear concentration calibration curves, the reader is referred to the previous section (ionic strength adjustors). With this technique ion measurements are as simple as pH measurements. It should be pointed out here, that one is far better off working at a single temperature and forgetting about the so-called automatic temperature compensation. As shown later in Chap. 6.2.1.2, this functions properly only in special cases.

With this quick and easy method analytical accuracies of about ±3% for monovalent and ±6% for divalent ions can be attained, provided there are no interferences of the first or second kind.

5.3 Titration Procedures for Determining Concentrations

5.3.1 Prerequisites

If concentration determinations are desired with accuracies of $< 0.5\%$, then titration procedures must be employed relying on ion-selective electrodes as end point indicators. At the basis of every titration is a stoichiometric reaction of the ion to be measured with a second substance present in the titration solution in precisely known amounts (= titer). If the stoichiometry of the reaction is known, the amount of measured ion in solution can be calculated from the amount of titrant needed to bring the reaction to completion. By measuring the amount of titrant with a precisely calibrated burette accuracies of better than 0.1% can be attained. In carrying this accuracy over into the total accuracy of the measurement, too careless assumptions are often made. *The accuracy of analysis is only comparable to the volumetric measurement accuracy if: a) the titer of the titrating solution can be determined correspondingly accurately, b) the stoichiometry of the titration reaction is known as accurately and remains constant, and c) the titration end point can also be determined with comparable accuracy!*

The first two conditions are more easily fulfilled. The value of a titration procedure rests on how accurately the end point can be determined. Visual recognition of titration end points (color change, end of precipitate formation) is prone to large subjective errors, so more objective methods have been sought for analytical chemical applications. Ion-selective electrodes provide such an objective alternative. These can be used in many ways for this application. If either the material to be determined or the titrant is directly indicated by an ion-selective electrode its usefulness is obvious (see Fig. 2). Even when none of the species participating in the titration reaction are directly indicated, one can still obtain a precise, electrode-indicated end point as follows: *A small amount of material which reacts with the titrating agent and which is itself specifically indicated by an ion-selective electrode is added to the sample solution.* In this way Ca^{2+} ions can be titrated with EDTA with the help of a Cu electrode and a trace of Cu^{2+} ions. Such a titration curve is illustrated in Fig. 44. Cu^{2+} forms a more

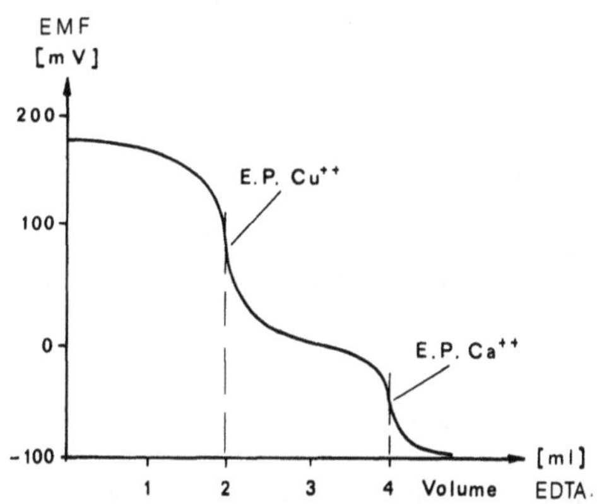

Fig. 44. Titration curve. (Equivalent amounts of Cu^{2+} and Ca^{2+} were titrated with EDTA using a Cu electrode as an end point indicator)

stable complex with EDTA (large complex formation constant) and is therefore complexed first as the complexing agent solution is added. As soon as the Cu^{2+} is complexed, the EDTA begins to react with the Ca^{2+} ions. The fact that the Cu electrode registers a further jump at the Ca^{2+} titration end point is because once all the Ca^{2+} has been complexed, the excess complexing agent can further reduce the Cu^{2+} activity according to the law of mass action. If a small amount of stoichiometrically bound Cu EDTA is added at the start of the titration, then naturally only one jump is seen. With this procedure the two complex formation constants should have very different values.

In titration procedures ion-selective electrodes serve only as end point indicators. In this case the analysis accuracy depends far less on an absolute and precise calibration of the electrochemical cell. Here one is far more interested in an exact determination of the *change* in the electrode EMF. It should be noted that only in symmetrical cases does the inflection point of the titration curve coincide with the end point of the reaction. In any reaction which is not 1:1 (such as $Ca^{2+} + 2\,F^- \rightleftharpoons CaF_2$ or $2\,Ag^+ + S^{2-} \rightleftharpoons Ag_2S$) the titration curve is not symmetric.

In precipitation titrations, the sharpness of the end point is determined by the solubility of the resulting precipitate. Attempts can be made to lower this solubility by going to non- or partially aqueous solutions. *Thus many titrations are carried out in ethanol, acetone or dioxane [such as the titration of F^- ions with a $La(NO_3)_3$ solution]. Cooling the solution can also lower the solubility.*

5.3.2 Titration Errors

It can be shown mathematically that considerable error results in asymmetric titration curves with poorly defined jumps if the inflection point (= maximum in the first derivative) is assumed to be the equivalence point [218]. Unfortunately, this is not taken into account often enough. The fact that many analysts have few problems with asymmetric titration curves plotted in this way, even though they have falsely indicated the end point, comes about because precisely the same mistake is made in determining the titer of the titrating solutions. The errors, in effect, cancel one another to a large extent. Also, with large, sharp jumps this error is normally negligibly small.

A further possibility for error when using ion-selective electrodes as end point indicators arises if the selectivity is insufficient. Considering that in this case the diluting effect on the interfering ion concentration will cause a change in EMF which has nothing to do with the titration reaction, it is easy to see its influence on a precise end point determination.

These errors arising from asymmetric titration curves, as well as from interfering ion indications were reckoned with by Carr [219, 220]. He started with the empirical form of the Nernst equation:

$$E = E' + S \cdot \log (f_m \cdot C_m + B) \tag{53}$$

where E, E' and S have the usual significance (Eq. 22), and B is the sum of the activities of the interfering ions also indicated:

$$B = \sum_{I} k^{pot}_{M-I} \, (f_I \cdot C_I)^{z_I/z_M} \tag{54}$$

where k^{pot}_{M-I} = selectivity coefficient measured ion – interfering ion

f_I = activity coefficient of the corresponding interfering ion

C_I = concentration of the corresponding interfering ion

z_I = charge of the interfering ion

z_M = charge of the measured ion.

He considered a precipitation titration:

$$nA + mT \rightleftharpoons A_nT_m \quad \text{(R35)}$$

and a complexometric titration:

$$A + T \rightleftharpoons AT. \quad \text{(R36)}$$

The magnitude of the end point jump depends on the equilibrium constant of the corresponding titration reaction, as well as on the concentration of the starting solution. Since the error in determining the end point of asymmetric titration curves becomes increasingly smaller with sharper end point jumps, both of these factors (starting solution concentration and equilibrium constant) will be considered as a function of various parameters. In a precipitation titration β is defined as:

$$\beta_P \equiv \frac{\upsilon \cdot K_S^{1/m}}{(C_{OS})^{\nu + 1}} \tag{55}$$

where υ = n/m, stoichiometry coefficient

K_S = solubility product of the precipitate $A_n\,T_m$

C_{OS} = starting concentration of the species A to be determined.

For a complexometric titration β_C is defined as:

$$\beta_C \equiv K_{CF} \cdot C_{OS} \tag{56}$$

where K_{CF} = complex formation constant of the species to be titrated with the complexing agent

C_{OS} = starting concentration of the species A to be determined.

The absolute magnitude of B is not so crucial for interfering ions also sensed by the electrode, but rather its magnitude relative to the concentration of the measured ion. Hence, to assess the error, an additional parameter b is defined:

$$b \equiv \frac{B}{f_m \cdot C_{OS}} \tag{57}$$

where B = same as in Eq. (54)

f_m = activity coefficient of the measured ion

C_{OS} = starting concentration of the measured ion.

In the original work, the influence of both of these normalized parameters β and b on analysis accuracy was investigated with the help of a computer, assuming that, as is usually the case, the inflection point in the titration curve plot is taken to be the end point of the chemical reaction. Symmetric precipitation titrations come out the best, showing no systematic errors even with relatively large β_P values of 0.1 to 0.01 and $b = 0$. For asymmetric titration curves, with $\upsilon = 2$ for example, an error of 55.3% results with $\beta_P = 0.1$ and $b = 0$.

To get a better feeling for these parameters, $\upsilon = 1, \beta_P = 10^{-6}$ and $b = 10^{-2}$ yield an error of 0.25%, larger than that of volumetric analysis. These theoretical errors were calculated without taking the dilution effect into consideration. *Taking this into account, in choosing a titration procedure with which one hopes to attain an error of better than 0.1%, one should make sure that* $\beta_P \leq 10^{-9}$ *or* $b \leq 3 \times 10^{-4}$. The errors in complexometric titrations, which are all asymmetric, are much larger. If an analysis accuracy of better than 0.1% is desired when using complexing agents in the titration, then one must be sure that $\beta_C \geq 10^4$ and $b \leq 10^{-2}$.

The error estimates above are all based on the assumption that both normalized parameters β and b are constant throughout the course of the entire titration. In practice it may happen that in some cases interfering ions are introduced into the sample solution via the titration (for example, in chelation titrations H^+ ions may be set free), thus causing b to increase. Anfält and Jagner [221] calculated the errors taking into account this non-constancy of β and b with the computer program "Haltafall" (IBM 360/65), available from the authors. For a titration using 100 ml of a solution with 0.01 M Ca^{2+}, 0.05 M Mg^{2+}, Cu^{2+}, Na^+ and 0.35 M NH_4^+ titrated with 0.1 M EGTA (ethylene glycol-bis-(β-aminoethylether)-N,N,N′,N′-tetraacetic acid), the second derivative method consistently indicated a systematic error of about −6%.

In addition, further errors can be made in titration procedures which are purely chemical or physico-chemical in origin. Supersaturation effects can sometimes be observed. This arises when for short periods of time the solution contains more ions than it should on the basis of the solubility product of the corresponding precipitate. *Precipitate formation, and thus the electrode EMF, is a time-dependent process* (such as the titration of traces of fluoride with a lanthanum solution, Fig. 50). *For this reason caution is required when working with automatic titration devices.*

A further complication of precipitation titrations is the occlusion or adsorption of excess ions present early in the titration (such as the argentometric halide determinations). With complexometric titrations the presence of other ions which also form complexes can interfere. Any or all of these influences may be the reason behind why the theoretically possible accuracy of better than 0.1% is not always attained. In such cases the reader is referred to the appropriate literature in the field of analytical chemistry.

These possible sources of error are listed here only so that the ion-selective electrodes are not alone held responsible for occasional incorrect results.

5.3.3 Sample Preparation for Titrations

The ion-selective electrode indicating the ion to be determined or an ion of the titration medium is attached to the indicating electrode terminal of the instrument and complemented with a suitable reference electrode. The sample solution concentrations should be such, that the β_P or β_C and b values described in the previous section are appropriate for the desired accuracy. If there is still some room for variation, the starting solution should be diluted as far as the β and b values will allow, in order to reduce the danger of occlusion and adsorption with precipitation titrations. At the same time, the ratio of the sample solution volume to the titrant solution volume should be chosen to be large to keep dilution effects as small as possible. With a typical starting volume of 100 ml, the titer of the titrating solution should be calculated so that less than 10 ml need be added. To minimize the effect of foreign electrolytes in the titrant on b (via the selectivity coefficients k_{M-I}^{pot}), high ionic strengths should be avoided, in contrast to direct potentiometry. Likewise any required pH's should be established with as little buffer material as possible. After these preliminary steps the electrodes are immersed in the sample solution and the titrant is slowly added with stirring. If an automatic titrator is not being used, then upon nearing the end point, the EMF changes are recorded after the addition of each 0.01 ml. The point of greatest change in EMF is taken as the end point. With all unsymmetric titration curves ($\upsilon \neq 1$), the titer of the volumetric solution should be determined with an accurately weighed sample of the measured ion. Concerning other end point indication techniques, the reader is referred to the literature [218].

Let us now consider a specific example for each type of titration:

Precipitation titration: Let us suppose our task is to determine to 0.1% the iodide content of an approximately 0.01 M I^- solution. We choose for this a precipitation titration with silver ions. According to the tables, the AgI produced in the reaction has a solubility product of about 10^{-16}. We then calculate the β_P value:

$$\beta_P = \frac{\upsilon \cdot K_S^{1/m}}{(C_{OS})^{\nu+1}} = \frac{1 \cdot (10^{-16})^{1/1}}{(10^{-2})^{1+1}} = 10^{-12}. \qquad (58)$$

According to the instructions in the last section, however, for an accuracy of 0.1% in the end result β_P need only be $\leq 10^{-9}$. In order to keep the amount of I^- occluded in the precipitate during the start of the titration small, we dilute our test solution 1:10 with distilled water. Now our starting concentration is 10^{-3} M, corresponding to a β_P value of:

$$\beta_P = \frac{10^{-16}}{(10^{-3})^2} = 10^{-10}. \qquad (59)$$

This value still assures a sharp end point jump in the titration curve. This same example for a chloride rather than iodide analysis would be far less favorable, since the solubility product of AgCl is only 10^{-10}. With a starting chloride ion concentration of 10^{-2} M, β_P is only 10^{-6}. This value would allow a final accuracy of 0.1% only under the most highly favorable conditions (extremely careful titer determination, using a standard having the same matrix as the sample solution and standardized analytical procedures). Here any diluting of the sample would smear out the end point.

Complexometric Titration: Suppose we are going to titrate Mg^{2+} ions with EGTA. From tabular complications we find that the Mg – EGTA complex formation constant is $K_{CF} \approx 10^5$. In order that the β_C value not fall below 10^4, then according to $\beta_C = K_{CF} \cdot C_{OS}$ the starting concentration of the Mg solution should be no lower than 0.1 M. If this concentration is substituted in the equation for b, then an interfering ion term of $B \leq 10^{-3}$ M must be attained if b is to remain $\leq 10^{-2}$. Only if these conditions are fulfilled can we expect an accuracy of 0.1%.

Naturally, all this holds only as long as the sampling, analytical procedure, titer determination, etc. do not introduce larger errors. *With not too dilute solutions, the titration method represents the most accurate analytical procedure employing ion-selective electrodes, since in this case neither the magnitude of the activity coefficient nor the liquid junction potential is important.* In addition, this technique allows the determination of a large number of ions, for which no selective electrodes yet exist. If a complexing agent is present in the sample solution, one can still obtain the total ion content of the solution with a complexometric titration, as long as the titrant complexing agent forms a more stable complex with the measured ion (larger K_{CF} value) than that originally present in the sample solution.

5.3.4 Titration to a Pre-determined EMF Value

5.3.4.1 On the Basis of a Titration Curve

As previously indicated, a more or less sharp jump in the titration curve occurs at the equivalence point of a precipitation or complexometric titration. With reasonably consistent samples, the inflection point appears at basically the same EMF value, even though the titration volumes may differ (corresponding to the different measured ion concentrations). Thus the "over-shoot" tail of the titration curve can be eliminated if the addition of titrating agent is stopped at the mV value of the equivalence point ("dead-stop" titration). The mV value is obtained by averaging the inflection points of a series of complete titration curves obtained using real sample solutions. In addition to this empirical procedure, the mV value can be ascertained with the help of a calibration curve for the corresponding electrode set-up and a knowledge of the equivalence concentration of the measured ion. Indeed, such a calibration curve should be used for testing the proper functioning of each electrode set-up.

The position of the titration curve prior to the equivalence point depends on the starting concentration. The course of the curve after the equivalence point depends on the corresponding equilibrium, and so is controlled by the solubility product K_S or the complex formation constant K_{CF}. Knowing the K_S or K_{CF} values, it is easy to calculate the corresponding equivalence point concentration. The final concentration of the complex is needed with complexometric titrations, but with precipitation titrations the value is fixed, since the precipitate forms a new phase. Locating this equivalence concentration on the calibration curve will yield the EMF value corresponding to the end of the titration. This technique of stopping a titration at a pre-determined EMF can be used to advantage with titrations in which kinetic factors lead to time-dependent EMFs (for example, supersaturation in precipitation titrations). In this case one waits for a stable EMF value to be established after each titrant addition before adding the next volume increment. In some cases the precipitation can be initiated by introducing a pure crystallite of the corresponding precipitate compound (more details can be found in quantitative analytical chemistry texts).

Difficulties arise with this method when dealing with broad end point jumps of variable b values of the sample solution (see Chap. 5.3.2). If the reaction kinetics are sufficiently rapid, this method can be easily adapted for an automated titration procedure.

5.3.4.2 On the Basis of a Concentration Cell Set-Up

In contrast to all other techniques, which use an ion-selective electrode and a reference electrode, for this technique a concentration cell must be built using two identical ion-selective electrodes or one ion-selective membrane and two identical reference electrodes. Fig. 45 shows a few possible concentration cell constructions. The principle of this method rests on the fact that the EMF of a concentration cell is zero if the measured ion activity is the same in both electrode compartments (Null point potentiometry). One can start with a measured ion activity of zero in the reference compartment (or more precisely, of an activity corresponding to the solubility of the active phase in this solution), and gradually increase the concentration of indicated ion through the precisely measured addition of a solution of the indicated ion of known concentration. This process is continued until a cell EMF of zero is obtained. At this point the known concentration of the reference compartment is identical to that in the adjacent sample solution. With this procedure no calibration of the electrodes is necessary. The slope of the Nernst response is also unimportant; it only affects the observed sensitivity of the EMF change per indicated ion addition. This technique is recommended if only a small amount of sample material is available, or with samples which cannot be contaminated. As Fig. 45B shows, even microliter samples can be measured accurately with this technique.

If one wants to carry out precise concentration determinations, a constant ionic strength must be established in both electrode compartments so that the same activity coefficients apply to both solutions. This also reduces the liquid junction potential at the contact point of the two solutions, further increasing the accuracy. If it is not possible to introduce any foreign substances into the sample solution (as in physiological measurements or with samples which are to be used further), then a capillary connection between the two solutions is used, and the final reference solution is given a composition similar to that of the sample matrix. In doubtful cases a non-destructive, non-polluting conductance measurement can be made on the sample solution and the same conductance is established in the reference solution. In favorable cases a number of

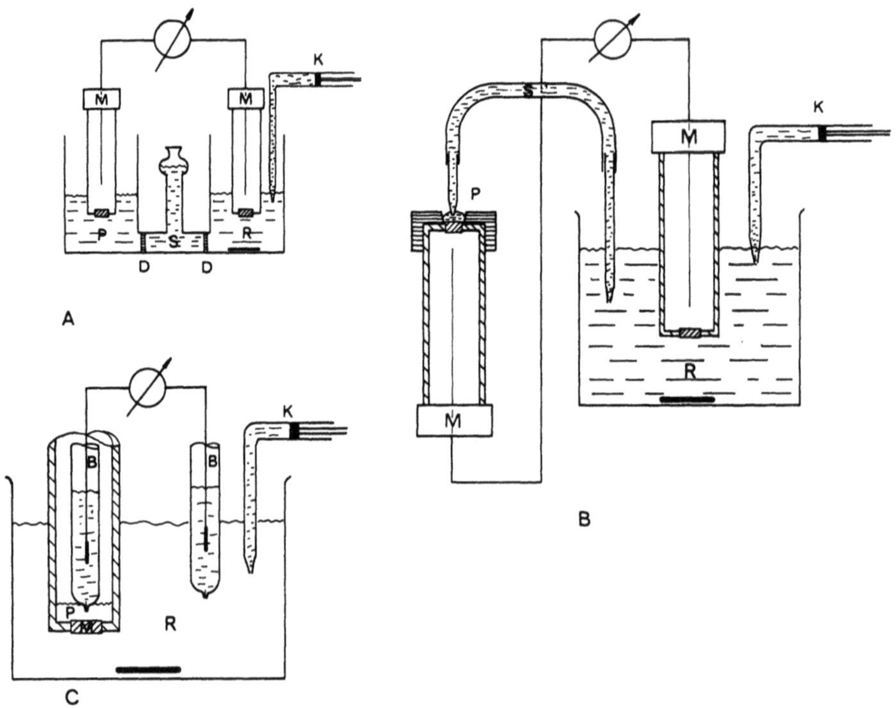

Fig. 45 A–C. Concentration cell constructions using ion-selective electrodes. **A:** Macro-construction with two identical ion-selective electrodes (*M*); the sample solution *(P)* is separated from the reference solution (*R*) by a salt bridge (*S*) and two diaphragms (*D*). **B:** Micro-construction with two identical ion-selective electrodes. **C:** Semimicro-construction with a dismountable ion-selective electrode and two reference electrodes (*B*)

old samples can be "purified" by selectively removing the indicated ion (through ion-exchange or precipitation), and the resulting solution used as a basis for the reference solution (this corresponds to using solutions titrated quantitatively to their equivalence points). If the sample solutions have very high concentrations of the indicated ion, then their dilution serves to increase the allowable concentration range between sample and reference solutions according to the considerations outlined in Chap. 5.1.2.

There are a great number of variations on this technique, but it is beyond the scope of this book to list them all here. Only the *inverse procedure* will be mentioned here, which starts with a reference solution having a higher concentration than the sample solution. Here the dilution necessary to achieve a zero EMF reading is measured, thus determining the sample concentration.

The zero EMF point can be ascertained graphically to achieve a more precise determination. For this the cell EMF after each incremental addition is plotted on semi-logarithmic paper (EMF vs. log $C_{\text{reference}}$). A linear titration curve is obtained with a slope of 59/z mV per power of ten at 25°C. The concentration of the sample can be read off of such a plot at the zero EMF point. This variation reduces the chances of error due to stray values (indicator fluctuations due to any of the interferences discussed in Chap. 4, careless reagent addition, reading errors, etc.) for purely statistical reasons (more measured values are utilized for such a plot). This method is especially recommended when no other similar samples are tested in order to be more certain of the analytical result.

Before starting such an experiment it is very important to test whether the EMF of the cell with the particular ion-selective electrodes used indeed indicates a zero mV value when the activities of the two adjacent solutions are equal. Thus at the beginning of a series of measurements one should determine as precisely as possible the EMF of the concentration cell when the same solution is placed in both cell compartments. As a rule this test usually results in a mV value very near zero. Only in cases where the two ion-selective electrodes exhibit different selectivity coefficients k_{M-I}^{pot}, and the interfering ion matrix deviates greatly among the various sample solutions does a variable "zero" value result. Thus this procedure serves to check the basis of this method of analysis. For the actual measurement, a suitable amount of sample solution for whatever construction is used, is placed in the sample chamber of the concentration cell, and the second ion-selective electrode is dipped in a sufficiently large beaker of the reference solution. As previously mentioned, the ionic strength of this solution should match that of the sample solution. To achieve the greatest accuracy possible, it is advantageous to establish the same, constant ionic strength in both solutions by means of an ionic strength adjustor. This also reduces the error caused by changes in the liquid junction potential at the contact zone between the two solutions. The calculation of the proper concentration of the reference solution can be simplified by using so large an initial volume of this solution, that the volume change due to standard solution addition can be ignored. This technique can be applied without concern when using a starting volume of 1 l and adding the standard solution with the aid of a precisely calibrated 1 ml piston burette. If the titer of the standard solution is chosen so that all samples can be titrated with less than 1 ml of solution, then the $< 0.1\%$ volume change can be neglected. In practice the reference solution should not have an initial concentration of zero for the reasons outlined in Chap. 1.3 (identity change of the active phase). With some ion-selective electrodes (glass and liquid membrane types) this procedure easily leads to potential drifting, especially when interfering ions are present in the matrix. Thus one starts with a precisely known concentration of the measured ion in the reference solution which is smaller than that expected in any of the samples. Titrations more accurate than ±0.1 mV of the preciously determined "zero" point are sometimes technically possible, but this only leads to a greater analysis accuracy if the variations in liquid junction potential between the two adjacent solutions are smaller than this value.

This null point technique also makes possible accurate analyses of very dilute solutions which cannot be obtained using direct titration procedures in which the measured ion activity of the sample solution itself is altered. Durst [225] was able to determine fluoride and silver ions in only 5–100 μl sample volumes with this technique. For amounts of material in the range 0.38 to 190 ng (!) fluoride and 0.054 to 54 μg silver observable errors of less than 1% were attained.

5.3.5 "Chemically Linearized" Titration Curves (One Point Titration)

Whereas titration curves with large, sharp, marked end point jumps are usually desired, with this technique one tries to attain the most linear relationship between EMF and volume (or concentration at constant volume) of titrant possible. Leithe [222] used a mixture of weak bases with overlapping pK values to linearize the titration curves of strong acids. The linearization represents nothing else but the superposition of the not very sharp titration curves of the individual bases (see Fig. 46). The advantage of this

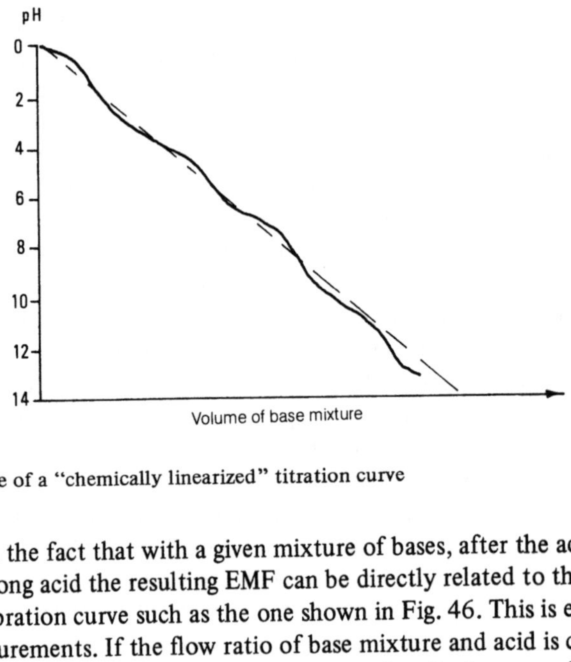

Fig. 46. Example of a "chemically linearized" titration curve

method lies in the fact that with a given mixture of bases, after the addition of a known amount of strong acid the resulting EMF can be directly related to the acid *concentration* via a calibration curve such as the one shown in Fig. 46. This is especially useful for flow measurements. If the flow ratio of base mixture and acid is constant, the EMF value provides a direct indication of the concentration. In the case of ion-selective electrodes, this method was predominantly developed by Oehme et al. [223] to the industrial stage. He reports, for example, a "direct potentiometric" determination procedure for sulfuric acid concentrations in anodizing electrolytes. Here the concentration range lay between 120–165 g/l and corresponded to an EMF change of only 8.4 mV when measured directly with a pH glass electrode. Merely by changing the concentrations of the bases used, an EMF change of 92 mV could be attained! [This increase in sensitivity is understandable if one considers that relatively large EMF changes accompany very small volume (concentration) changes near the end point of a titration curve.]

Since the complexation of Al^{3+} ions by F^- procedes through AlF^{2+} (pK~6), AlF_2^+ (pK~11), AlF_3 (pK~15), AlF_4^- (pK~18), AlF_5^{2-} (pK~19) and AlF_6^{3-} (pK~20), no sharp jump can be expected during a titration of fluoride with Al^{3+} ions. Using this technique, Oehme and Dolezalova [224] were able to determine Al^{3+} concentrations in the range 10–15 g/l with the help of a fluoride electrode. Using a conditioning solution containing 2 M acetate buffer and 10^{-3} M F^-, they obtained a 32 mV EMF change for this small concentration range relative to the 3.4 mV theoretically expected for a direct indication of trivalent ions. This clearly illustrates that higher valent ions can be determined with sufficient accuracy using this technique in spite of their small Nernst factors. According to these authors, other fluoride complex-forming species such as Fe^{3+}, Th^{4+}, Zr^{4+} and UO_2^{2+} can be determined this way, as well as Ni^{2+}, Fe^{2+} and Au^+ with CN^- as "helping ion", Hg^{2+} with Br^-, SCN^- with Cu^{2+} and $S_2O_3^{2-}$ with Ag^+.

5.4 Concentration Determinations with the Help of a Standard Addition with a Known Electrode Slope S

5.4.1 Measuring the Change in EMF upon Addition of a Standard Solution to the Sample Solution

Fundamental to this procedure is the inference which can be made about the starting concentration of a precisely measured amount of sample solution from the observed change in EMF upon addition of a precisely measured amount of standard solution containing the appropriate indicated ion.

In order to derive the formula to be used, one starts with the empirical form of the Nernst equation:

$$E_1 = E' \pm S \cdot \log (C_X \cdot f_X \cdot k_X) + \Delta E_{LJ} \tag{60}$$

where:
- C_X = unknown concentration of the measured ion in the sample
- f_X = activity coefficient of the measured ion in the sample
- k_X = fraction of free measured ion in samples containing complexing agents
- ΔE_{LJ} = variations in the liquid junction potential at the reference electrode electrolyte/sample solution phase boundary.

After addition of a certain amount of a standard solution and raising the concentration of the measured ion by an amount ΔC, the cell EMF is:

$$E_2 = E' \pm S \cdot \log (C_X + \Delta C) f'_X k'_X + \Delta E_{LJ}. \tag{61}$$

If the amount of added standard solution is kept small enough so that the total ionic strength of the solution is not substantially altered, and so that any complexing agents present in the sample solution completely complex this added amount of measured ion, then $f_X \cdot k_X = f'_X \cdot k'_X$. In addition, E' and ΔE_{LJ} will remain constant and fall out of the difference $E_2 - E_1$, so that:

$$\Delta E = E_2 - E_1 = S \cdot \log \frac{C_x + \Delta C}{C_x} \tag{62}$$

which upon rearrangement gives:

$$C_x = \Delta C \, (10^{\Delta E/S} - 1)^{-1}. \tag{63}$$

If a volume V_s of standard solution with concentration C_s is added to a volume V_t of the test solution, then, ignoring dilution effects:

$$C_x = C_s \left(\frac{V_s}{V_t}\right) (10^{\Delta E/S} - 1)^{-1}. \tag{64}$$

Including the dilution effect this becomes:

$$C_x = C_s \left(\frac{V_s}{V_t + V_s}\right) \left[10^{\Delta E/S} - \left(\frac{V_t}{V_t + V_s}\right)\right]^{-1}. \tag{65}$$

This method of evaluation can also be used when the concentration of the measured ion in the test solution is reduced by a precisely known amount (preciptitate or complexation, subtraction technique) rather than increased.

The magnitude of the variables C_s, V_t, V_s and S in Eq. (65) can usually be determined very accurately. The largest error arises in determining ΔE. This involves a small difference between two large numbers, which from a mathematical standpoint can easily become very imprecise. On the other hand, there are limits to the size of ΔE. The ionic strength and degree of complexation in the sample solution must not be changed substantially upon standard addition. With larger additions it is also not clear that one is working in a region with the same slope. Nonetheless this procedure is often employed, because it is one of the few methods, *which under certain conditions detects the concentration of complex-bound material as well.* Complex binding of measured ions is a frequent occurrence. It can be discovered through a discrepancy between analytical data obtained using two different evaluating methods employing ion-selective electrodes. As an example, the determination of fluoride in sea water using a concentration calibration curve yields 0.6 ppm, while the addition method indicated a value of 1.2 ppm [225]. This implies that half of the fluoride in sea water is bound up in complexes (e.g. with the magnesium ion). With regard to physiological measurements it is important to realize that a fraction of the ion to be measured can be bound to proteins [226].

To assess the possibility of determining complex-bound ions also with this technique, let us briefly refer back to the chemistry of complex formation. In a complexation reaction:

$$\mathrm{A + T \rightleftharpoons AT} \tag{R37}$$

the complexing agent T forms a complex AT with the indicated ion A. Hence the concentration of free A ions is diminished. As in all chemical reactions an equilibrium is established, but in the case of strong complexing agents it lies entirely on the right side. Applying the law of mass action to this reaction yields:

$$\frac{a_{AT}}{a_A \cdot a_T} = K_{CF} \tag{66}$$

where K_{CF} is the complex formation constant. As a rule these are very large (10^3 to 10^{12}), so with an excess of complexing agent the concentration of free A ions is very small. At large and constant excess of T, however, it is proportional to the starting amount of A, since the constant factor a_T can be combined with K_{CF}. Pictorially (see Fig. 47), this falls in a region of the titration curve which could be called the

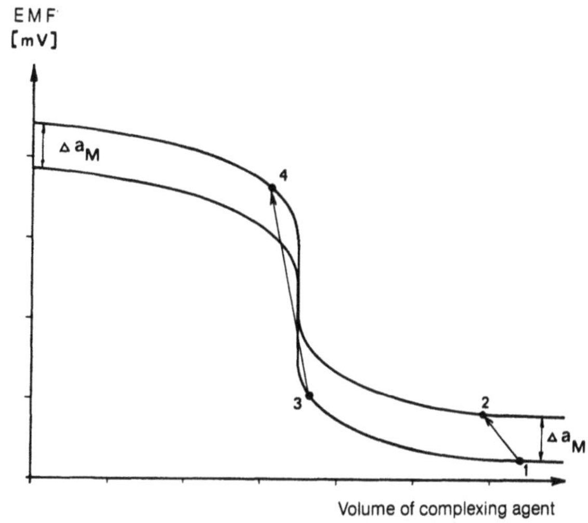

Fig. 47. Working points on the virtual complexometric titration curve for the standard addition technique; points 1 → 2 yield correct determinations, points 3 → 4 give erroneous results

post-equivalence region with an excess of titrating agent. In contrast to precipitation titrations in which a new phase is formed, in complexometric titrations the EMF value depends, in this region, on the starting concentration of the solution. Here, however, the indicated ion concentration is smaller by the factor K_{CF} multiplied by the excess of complexing agent. This is the physical chemical basis for being able to determine the measured ion concentration without regard to the complexing agents present using this method. Introducing an amount of measured ion which would raise its concentration from 1 M to 2 M if the complexing agent were not present, in fact only changes the concentration from, for example, 1×10^{-8} to 2×10^{-8} M. However, since both these changes have the same ratio, the constant complex formation constant factor is unnoticed in the measurement of ΔE; only the absolute magnitude of the cell EMF is smaller. It is now easy to see why a large excess of complexing agent is required in the sample solution. Otherwise the ratio AT/A would not remain constant. In this case one would be working in the end point jump region of the titration curve (see Fig. 47, points 3 → 4). In such a region the prerequisites of this procedure are not at all fulfilled. This is to be kept in mind when trying to use this method to investigate unknown systems. *In natural systems (physiology) such a large excess of complexing agent is not always present. In many instances one is working in a region near the equivalence point of a complexometric titration.* This may help to clarify some of the discrepancies which come up from time to time in the analyses of these types of solutions.

For this procedure the slope S of the electrochemical cell is needed *within the concentration region in question.* This should be determined before starting any measurements through the construction of a concentration calibration curve. The slope usually remains fairly constant. We have seen in the last section how important a constant ionic strength is for an accurate analysis, and if a complexing agent is present in the sample solution, a large excess of this agent. For this reason an ionic strength of about 1 M is established in the sample solution by dilution with an ionic strength adjustor (see Sect. 5.1.2). If one is uncertain whether a given complexing agent is present in the sample solution in sufficient excess, one of two procedures may be chosen. A species may be added which forms a more stable complex with this agent (displacement reaction), thus setting the measured ion free (such as AlF_6^{3-} + citrate). Alternatively, a stronger complexing agent can be added to the sample solution in sufficient excess before the measurement, as long as the measured ion concentration does not fall below the detection limit. In this case both additions can be combined, and the sample solution is diluted 1:1 with this conditioning solution (often 5 M $NaNO_3$ + 1 M complexing agent). In such situations the calibration curve to determine the slope is constructed from data obtained using the same conditioning solution dilution procedure. According to the content of the sample solution, the concentration of the measured ion in the standard solution is chosen so that an addition of about 10 ml of this solution to the sample solution causes a change in EMF of about 30 mV for monovalent ions and about 10 mV for divalent ions. If a starting volume of 10 ml of sample solution is diluted with 10 ml of conditioning solution, the standard solution should be about 5 times as concentrated as the original sample solution. *If exactly 10 ml of the standard solution are added, then a complete analysis can be carried out with one standard solution, one conditioning solution and one pipette.* This simplicity is, naturally, very important for field measurements. An accuracy of about 1% can be achieved with this method.

5.4.2 Measuring the Change in EMF upon Addition of the Sample Solution to a Standard Solution

This is precisely the reverse of the previously described technique. Here a certain amount of sample solution is added to a known volume of standard solution of the appropriate indicated ion in which the electrodes are immersed, and the original measured ion concentration of the sample is calculated according to the following formula:

$$C_X = C_S \left[\left(\frac{V_S + V_t}{V_t} \right) \cdot 10^{\Delta E/S} - \left(\frac{V_S}{V_t} \right) \right] . \qquad (67)$$

The symbols have the same meaning here as in Eq. (65). The same assumptions made in the previous derivation are also valid here, namely $f_X \cdot k_X = f'_X \cdot k'_X$ and E' and ΔE_{LJ} are constant.

One advantage of this variation is that the minimum volume needed for the immersion of the two electrodes is provided by the standard solution. If the measured ion concentration in the standard solution is adjusted to be 100 times smaller than that expected in the sample solution, then a very small volume of sample solution suffices to produce an optimal EMF change (about 30 mV for monovalent and 10 mV for divalent ions). If, at the same time, the standard solution exhibits a sufficiently high total ionic strength, then the activity coefficient is hardly changed by the addition of a small test solution volume, so the conditions for a direct concentration measurement are fulfilled. *Caution is also required here if the sample matrix contains complexing agents. For the reasons previously stated, one must be sure that even after dilution with the standard solution, the complexing agent is still present in large excess.* Otherwise the experiment is no longer being carried out at the proper position (Fig. 47 points 1 → 2) on the virtual titration curve and the assumptions made in the derivation of Eq. (67) no longer hold. Here again, there are two possible ways to avoid this difficulty. Either an ion which forms stronger complexes with the complexing agent present in the sample solution than does the measured ion (for example, citrate binds Al^{3+} ions, setting F^- free) is added to the standard solution at the start, or a stronger complexing agent is added in excess (citrate usually forms stronger complexes than acetate). The difference between these two procedures is that the first sets measured ions free, resulting in a relatively high concentration, whereas in the second case this concentration is reduced by a constant factor according to the complex formation constant. *Since this factor can often take on values of about 10^{-10}, problems associated with detection limits may arise with this method. This comes about especially if the first term in the empirical Nernst equation expression ($a_M + k_{M-I}^{pot} \cdot a_I$) is so small that the interfering ion term is potential-determining. This holds for the previous addition technique as well.*

This second variation of the addition method is also more convenient, because the necessary conditioning agents (ionic strength or pH regulators, complex-forming ions or, when absolutely necessary, complexing agents) can be easily included in the premixed standard solution. Of course, this method can also detect species for which no ion-selective electrodes are available. The only requirement is that such species *react stoichiometrically* with some ion which can be directly indicated. Here also the change in concentration of the standard solution upon addition of the sample solution results in a corresponding change in EMF. Minding the proper signs, Eq. (67) can be used to calculate the desired concentration here as well.

When uncertain whether the conditions $f_X \cdot k_X = f'_X \cdot k'_X$ as well as E' and ΔE_{LJ} = const. are fulfilled, an empirical test is recommended. To this end, a volume of water corresponding to the volume of sample solution to be used later is added to the prepared standard solution (with or without conditioning solution), and the resulting EMF change α is observed. This EMF change, naturally, arises from the dilution effect. With this correction factor, a more general expression can be derived [227]:

$$\Delta E = \alpha + S \cdot \log \left[1 \pm \frac{V_t \cdot C_X}{V_s \cdot C_s} \right] . \tag{68}$$

The minus sign here applies to the cases in which the species to be determined lowers the standard solution concentration. In this equation the empirically determined α term represents the following:

$$\alpha = S \cdot \log \left[\frac{V_s}{V_s + V_t} \cdot \frac{f'_X \cdot k'_X}{f_X \cdot k_X} \right] + \Delta E_{LJ} . \tag{69}$$

According to Eq. (68), the desired sample concentration C_X can be expressed:

$$C_X = \pm \frac{C_s \cdot V_s}{V_t} \left[10^{(\Delta E - \alpha)/S} - 1 \right] . \tag{70}$$

This parameter α, empirically determined from a series of measurements, corresponds to a correction of the measured EMF change. The $\Delta E - \alpha \equiv \Delta E_C$ term can be thought of as a corrected ΔE value. If the conditions mentioned above have been met, then α has a value near 0 mV. Orion has published in tabular form the values corresponding to the term in brackets in Eq. (70) [$1/(\Delta C/C_k)$, or f factor in their notation], calculated from the corresponding ΔE_C values (*assuming the cell exhibits the Nernst slope*). These values, effectively the changes in concentration upon addition, can be found in the Appendix. To calculate the starting sample solution concentration, the appropriate f value is taken from the table and multiplied by $\frac{V_s \cdot C_s}{V_t}$. This yields the concentration of sample solution in the same units as those used to express the standard solution concentration (M, ppm, meq., etc.).

5.5 Concentration Determinations with the Help of a Standard Addition with an Unknown Electrode Slope S

The empirical form of the Nernst Equation

$$E = E' \pm S \cdot \log a_M \tag{71}$$

now contains two unknowns, E' and S. Two equations are required to determine two unknowns. For electrode standardization procedures this means that E' and S can be determined only through the measurement of two different standard solutions of known concentration. In the cases of the addition methods described in the last section this can be done very elegantly with a double standard addition procedure. A single standard addition followed by a known dilution also provides the same information.

5.5.1 Method of Double Standard Addition

To derive the evaluating formula we again start with Eq. (60). An ion-selective electrode and a reference electrode dipping in a solution of unknown measured ion concentration C_X will exhibit an EMF according to:

$$E_1 = E' + S \cdot \log\,[f_X \cdot k_X \cdot C_X] + \Delta E_{LJ}. \tag{72}$$

The symbols are defined in Eq. (60).

Now a precisely known volume of a standard solution containing an accurately established measured ion concentration is added. Due to the concentration rise ΔC, the EMF reaches a new value E_2:

$$E_2 = E' \pm S \cdot \log\,[f'_X \cdot k'_X \cdot (C_X + \Delta C)] + \Delta E_{LJ}. \tag{73}$$

So far this resembles the addition processes described above. Here, however, after noting this new EMF, a second equal portion of standard measured ion solution is added to the sample solution. If volume changes can be neglected, the final EMF is given by:

$$E_3 = E' \pm S \cdot \log\,[f''_X \cdot k''_X \cdot (C_X + 2\,\Delta C)] + \Delta E_{LJ}. \tag{74}$$

If $f_X \cdot k_X = f'_X \cdot k'_X = f''_X \cdot k''_X$ and E' and ΔE_{LJ} are constant through both EMF changes, then:

$$\Delta E_2 = E_2 - E_1 = S \cdot \log \frac{C_X + \Delta C}{C_X} \tag{75}$$

$$\Delta E_3 = E_3 - E_1 = S \cdot \log \frac{C_X + 2\,\Delta C}{C_X}\,. \tag{76}$$

Dividing ΔE_3 by ΔE_2 yields:

$$R = \frac{\Delta E_3}{\Delta E_2} = \frac{\log\left(\dfrac{C_X + 2\,\Delta C}{C_X}\right)}{\log\left(\dfrac{C_X + \Delta C}{C_X}\right)}, \tag{77}$$

and the constant factor S drops out. At the same time we have arrived at an equation with the desired relationship between the EMF change and the corresponding change in concentration. Since it is difficult to solve this equation explicitly for C_X, a number of R values for various $C_X/\Delta C$ ratios are listed in the Appendix [227]. Use of these values simplifies the calculation if even concentration units are present in the added solution (such as 1 ppm, 1 meq/l, 10 meq, 1 x 10^{-3} M, etc.).

Naturally, with this double standard addition procedure the same precautions (constant ionic strength, constant degree of complexation, ΔE_{LJ} = const.) must be observed as with the simpler addition procedures. *Since the electrode slope is calibrated within a relatively small concentration range (2 ΔC), it is possible with this method to work in regions where the calibration curve is somewhat bent.* There is an advantage when working near the detection limit.

5.5.2 Method of Standard Addition with Subsequent Dilution

With the double standard addition method it is not the electrode slope precisely at the sample concentration which is determined, but rather the slope in a somewhat more concentrated region (between $C_x + \Delta C$ and $C_x + 2\,\Delta C$). By diluting the solution after the first addition, a slope can be obtained much nearer to the sample concentration. In this procedure, as usual, a known amount of standard measured ion solution is added and the EMF change ΔE_{ad} is noted. For this a concentration change ΔC approximately that of the sample solution is chosen. To position the working point precisely in the middle, the solution is then diluted 1:1. The EMF change due to this dilution ΔE_d is also noted. The slope of the electrode used can then be calculated with the following equation:

$$S = \frac{\Delta E_d}{\log(V_{final}/V_{initial})} = \frac{\Delta E_d}{\log 2} = \frac{\Delta E_d}{0.301}. \tag{78}$$

Substituting this in Eq. (63) yields:

with $$R = \Delta E_{ad} / \Delta E_d \tag{79}$$

$$C_x = C(10^{R \cdot 0.301} - 1)^{-1}; \tag{80}$$

In Appendix 6 the $C_x/\Delta C$ values corresponding to various to-be-determined R values are tabulated. The evaluation is even easier if the expression $(10^{R \cdot 0.301} - 1)^{-1} \equiv A$ is tabulated as an increment factor depending on E_{ad} and E_d (see Appendix 7).

The starting point for this treatment was again Eq. (63), which was derived ignoring volume changes. This means that the sample volume must be kept large relative to the volume of added standard solution. This can be accomplished by making the standard solution to be added at least 100 times as concentrated as the sample solution. Then in order to double the sample concentration less then 1/100 the sample volume is needed. With sample solutions of low ionic strength ($< 10^{-2}$ M) water can be used as the diluting medium. If the sample solution has a higher ionic strength, then the change in activity coefficient accompanying a 50% concentration change is too large, and a diluting solution must be used with a concentration of inert electrolyte similar to the ionic strength of the sample solution. (In doubtful cases conductance measurements should be made to estimate the ionic strength of the sample solution.) The same holds true if the indicated ion must be set free through the use of a complexing agent. In this case the diluting solution should contain the same agent in the same concentration as the sample solution.

The main advantage of this technique lies with quick survey measurements near the detection limit. Using this technique 10^{-8} M fluoride solutions could be analyzed with 30% accuracy. This is quite noteworthy in view of the fact that the calibration curve begins to derivate from linearity with the fluoride electrode at 10^{-6} M – two orders of magnitude higher [228].

The problem with this method is, again, the fact that a small difference ΔE must be measured between two relatively large numbers, E_2 and E_1. Since each EMF measurement has its own uncertainty, ΔE can become quite inaccurate, especially when its value is small. From the rules of error propagation it is clear that the individual EMF measurement errors are additive in calculating ΔE. According to the considerations in Chap. 2, when working with cells with liquid junction, accuracies of ±0.1 mV must be dealt with. This means a minimum error in ΔE of ±0.2 mV. Unfortunately, as explained in Sect. 5.4.1, ΔE cannot be made large enough that this error is negligible by increasing the size of the standard addition. Carrying these uncertainties over into the final concentration values, for monovalent ions (the most favorable case) direct potentiometric analysis yields an error of at least 0.8%; simple addition methods with known electrode slopes at least 1.6%, and in this last instance of unknown slope and a dependence on two EMF measurements, at least 8%. For divalent ions these errors are all twice as great.

5.6 Practical Example of the Addition Method: Sodium and Potassium Determination in Blood Serum

Required Apparatus:

1 pH meter (precision of ±0.1 mV)

1 Sodium-Selective Glass Membrane Electrode

1 Potassium-Selective Ion-Carrier Electrode

1 Ag/AgCl Reference Electrode with Sleeve Diaphragm and Double Salt Bridge

1 Thermostated Sample Cell

Preparation:

In normal blood serum one can expect a sodium content of about 140 meq/l and a potassium content of about 5 meq/l. Due to the large excess of sodium ions and the small k^{pot}_{Na-K} selectivity coefficient of the glass membrane electrode, there is no need to worry about direct interference from potassium ion indication. The potassium ion electrode exhibits a k^{pot}_{K-Na} of 10^{-4}. This corresponds to a value of 0.014 meq/l for the $k^{pot}_{M-I} \cdot a_I$ term in the empirical Nernst equation, which is negligible compared to the 5.0 meq/l potassium activity (a_M) present.

Since we do not know the state of Na or K in blood serum, we choose an addition procedure. Let us suppose our desired accuracy is 2%, so we select a method making use of a known electrode slope and employ a reference electrode with a sleeve diaphragm. From the considerations in Chap. 2.3 we know that this cell construction provides a very stable liquid junction potential. Since this type of diaphragm is characterized by a large electrolyte outflow, the outer reference electrolyte should not contain Na^+ or K^+ ions. For this reason a 1 M lithium trichloroacetate electrolyte is recommended for filling the outer salt bridge compartment [229]. This can be prepared by neutralizing a 2 M LiOH solution with 2 M trichloroacetic acid.

We wish to dilute the serum sample as far as the linear region of a previously determined calibration curve will allow. In this way we hope to work in a region of low ionic strength where the influence of ionic strength on activity coefficients is not so pronounced (see Appendix 1). We choose to do it without the addition of an ionic strength regulator in order to avoid the possibility of salting out the protein molecules, which could occlude some portion of the measured ion.

Since 1 meq/l corresponds to 10^{-3} N, we prepare a series of standard solutions starting with 0,1 N NaCl and KCl solutions and decreasing in concentration from there (solutions under 10^{-4} M must be freshly prepared for each measurement). With the help of these solutions and the Ag/AgCl double junction reference electrode we then establish the calibration curves for each of the ion-selective electrodes. This determines the characteristic working range (linear calibration curve region) of each electrode. In addition, from the EMF change in the lowest concentration region where the curve still progresses linearly, we can determine the electrode slopes (S = EMF/power of 10 concentration change) of the respective indicating electrodes. From experience, one can expect to find a slope of 50–57 mV per power of ten at 25°C for both of these electrode types.

This is also the best way to check for proper functioning of the electrodes. This should be done with every ion-selective electrode after a prolonged period of non-use. In addition, the electrode slopes obtained with this averaging graphical method are more accurate than those determined by the procedures described in Sect. 5.5.1 and 5.5.2. If one has previously carried out this approximately half-hour long procedure, it is at once obvious if the potassium-selective ion solvating electrode must be regenerated. The calibration curve also shows us how much room we have to play around with the dilution of the serum sample. With intact indicating electrodes, we will find that the serum can be diluted 1:100 for the sodium determination, but only 1:10 for the potassium measurement without entering a non-linear region of the calibration curve. Now that this is clear, let us turn our attention to the optimum concentration of the standard solution to be added:

After dilution we must reckon with 1.5 meq/l Na and 0.5 meq/l K concentrations. In order to avoid volume corrections, we choose over 100-fold more concentrated standard solutions. For the sodium determination we choose a 200 meq/l Na solution, and for potassium a 100 meq/l K solution.

Since the added volume of these standard solutions must be measured very accurately (naturally, any error here contributes to the total error), they are best loaded into precisely calibrated 1 ml volume micrometer burettes. For the volume of 0.5 ml of this solution corresponding to the desired optimum EMF change of 30 mV, the accuracy of the micrometer burette is ±0.003 ml. In order that this full accuracy be attained in reality, care should be taken that the tip of the burette is drawn to a fine capillary, and only dips in the sample solution during the actual addition process. This completes the preparatory work. The actual analysis procedure is much shorter, so that especially with multiple analyses, the electrochemical procedures are used to full advantage.

Analysis Procedure:

5 ml serum sample is diluted
↓ with distilled water
50 ml volumetric flask

Na branch:

10 ml are pipetted for Na determination and diluted with distilled water
↓
100 ml volumetric flask
↓
30 ml are pipetted and placed in a 100 ml beaker with magnetic stirrer, electrodes are immersed and stirring is commenced
↓
1st mV value recorded = E_{Na1}
sodium solution added via burette tip immersed in sample solution until EMF~30mV more positive
added volume = V_{SNa} recorded
↓
2nd mV value recorded = E_{Na2}

K branch:

30 ml are pipetted for direct K determination, placed in a 100 ml beaker with magnetic stirrer, electrodes are immersed and stirring commenced
↓
1st mV value recorded = E_{K1}
potassium solution added via burette tip immersed in sample solution until EMF~30 mV more positive
added volume = V_{SK} recorded
↓
2nd mV value recorded = E_{K2}

Concentration calculations:

$$E_{Na2} - E_{Na1} = \Delta E_{Na} \tag{81}$$

$$E_{K2} - E_{K1} = \Delta E_{K} \tag{82}$$

$$\text{meq Na/l} = 100 \cdot (\text{meq Na/l})_{\text{standard}} \cdot \frac{V_{SNa} \text{ in ml}}{30} \cdot [10^{\Delta E_{Na}/S_{Na}} - 1]^{-1} \tag{83}$$

$$\text{meq K/l} = 10 \cdot (\text{meq K/l})_{\text{standard}} \cdot \frac{V_{SK} \text{ in ml}}{30} \cdot [10^{\Delta E_{K}/S_{K}} - 1]^{-1}. \tag{84}$$

5.7 Concentration Determinations with the Help of a "Mathematically Linearized" Titration Curve

With sufficiently large starting concentrations, the most accurate analyses (~0.1%) can be achieved with ion-selective electrodes by means of a quantitative and stoichiometric titration. With smaller starting concentrations the end point determination becomes more difficult; in addition, the systematic errors outlined in Sect. 5.3.2 also arise in the case of poorly-defined, asymmetric titration curves or if interfering ions are also indicated. By monitoring the titrated ion with an ion-selective electrode, the decrease in ion concentration due to the progressing titration can easily be followed until about 90 to 95% of the equivalence point volume. The closer the equivalence point is approached, however, the lower the concentration of free measured ion in solution. At this point the first interfering terms begin to become important: At sufficiently small measured ion concentrations, some interfering ions begin to increasingly determine the electrode potential. At the same time, passing the equivalence point in the case of asymmetric titrations ($\upsilon \neq 1$) causes the mathematical relationship between the electrode potential and the added volume of titrant to change from what it was near the beginning of the titration. In spite of this, the usual evaluation procedures start out by assuming that the data on both sides of the turning point can be described by a well-defined, homogeneous mathematical function. Since this is unfortunately not the case in most real situations, the systematic errors discussed in Chap. 5.3.2 arise. One solution to this problem would be to use only that portion of the titration curve jump where it is clear that a well-defined mathematical relationship exists between the electrode potential and the measured ion concentration. Approaching the end point from either side, its exact location must then be determined by extrapolation. In this case the traditional *titration curve must be linearized.* Gran [230] reports a procedure for determining the equivalence point of potentiometrically indicated titrations by means of extrapolating a linearized titration curve. Unfortunately this procedure has been largely overlooked to date. It offers a number of advantages and has fewer drawbacks than any of the previously discussed procedures. The somewhat more involved calculations need no longer be a discouraging factor thanks to the inexpensive pocket calculators available today.

Gran's method is based on the idea that a linear titration curve is obtained if the concentration of the indicated ion is plotted rather than the corresponding electrode EMF. This is due to the simple relationship: Titration with 10% of the equivalence volume = 10% less free ion in solution; 50% equivalence volume titrated = 50% less; 100% titrated = no free measured ion in solution. With titrations followed by conductance measurements a direct proportionality (not logarithmic, as with EMF measurements) is obtained between the measured ion concentration and the conductance, resulting in linear concentration curves which can be extrapolated. To get the concentration corresponding to a particular EMF one can either use a suitable calibration curve, or read the concentrations directly off of a logarithmically divided instrument scale. The empirical Nernst equation can also be mathematically manipulated into a suitable form:

$$E = E' + S \cdot \log a_M \quad (85)$$

$$E/S = E'/S + \log a_M \quad (86)$$

$$10^{E/S} = 10^{E'/S} + a_M \quad (87)$$

$$\text{antilog}\, E/S = \text{const.} + a_M . \quad (88)$$

One can work with this form of the Nernst equation as well. Fig. 48 shows an electrode calibration curve obtained by plotting the antilog of the electrode EMF divided by the

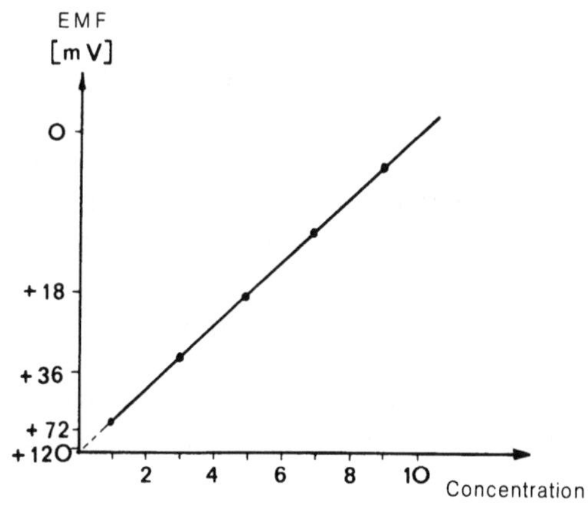

Fig. 48. Calibration curve for a monovalent anion according to the Gran method

slope as ordinate against the concentration values on a linear abscissa scale. This illustrates one property of such a plotting procedure; namely, that the calibration curve can be extrapolated to zero activity or zero concentration where the horizontal axis is intersected. One can easily titrate to this point experimentally, but the correct EMF at this point is difficult to determine. These difficulties arise due to the uncertain factors which cause a potential to be established in a solution containing no measured ion. In such solutions EMF drifts are observed until the partition equilibrium of the measured ion between the active phase and solution is established. At this point the measured ion concentration is, naturally, no longer zero. In cases where the solubility of the active phase is extremely small (as with Ag_2S where $K_{sp} \sim 10^{-51}$), impurities determine the electrode potential and irreproducible EMF values are obtained. A zero concentration of the measured ion at the equivalence point can hardly be achieved or measured in a real titration.

Nonetheless, according to the stoichiometric calculation, the titration is considered as being 100% quantitative. Thus the extrapolation to a zero measured ion concentration used here represents the first method which actually fulfills the conditions assumed in the calculation. Since the slope S is constant, the actual EMF values in mV can also be plotted on special graph paper instead of the corresponding $10^{E/S}$ values. Then the calibration curve graphically displays the relationship between EMF and concentration, except that in this case the concentration scale is linear while the EMF scale increases exponentially. The extrapolation to zero concentration intersects the ordinate at a mV value corresponding to the additive constant $10^{E'/S}$.

Such a calibration curve is easy to construct. The indicating and reference electrodes are immersed in 1 l of conditioning solution and successive 1 ml portions of a solution 100 times more concentrated than the expected sample concentrations to be measured are added via a precisely calibrated 10 ml burette. In this way a calibration curve for solutions exhibiting 100-fold lower measured ion concentrations is obtained. Concentration changes due to dilution effects can be neglected in this case. For situations in which they cannot be neglected, the expression $(V_0 + V)\,10^{E/S}$ must be plotted as the ordinate. Properly calibrated graph paper [231] (ordinate: $(V_0 + 10\%)\,10^{E/58}$, abscissa: volumes 0 to 10 ml) and rulers [232] are already commercially available. With these a 10% volume change caused by standard addition is taken into account. Hence these correct for a 100 ml starting solution to which 0–10 ml standard solution are added. Here, however, we have already arrived at an important application. What happens if such a calibration curve is constructed from a sample solution already containing some indicated ion? Then the plotted term $10^{E/S}$ is composed of:

$$10^{E/S} = \text{Const.} + a_{\text{sample}} + a_{\text{standard}}. \tag{89}$$

In this case one finds a parallel displacement of the calibration curve previously obtained with a solution containing no measured ion (Fig. 49). In order to reproduce

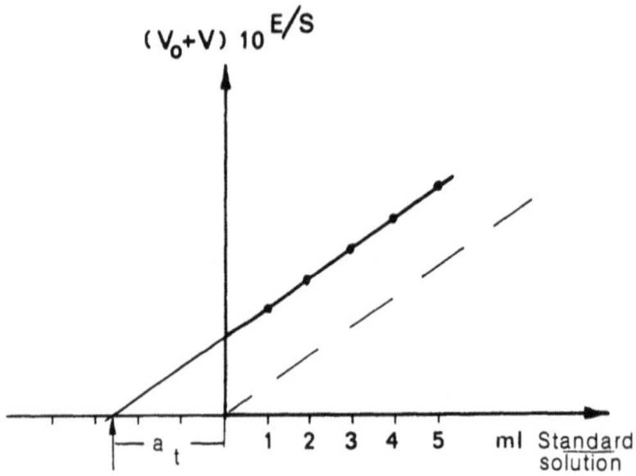

Fig. 49. Standard addition technique according to Gran

the original calibration curve, this new line must be displaced by an amount a_t along the activity axis (or concentration axis for constant activity coefficients). This can only mean, however, that the measured ion activity in the original test solution is precisely a_t.

One can also imagine the reverse situation, where a material which reacts quantitatively with the measured ion is present in the sample solution, thus removing measured ions according to the solubility of the resulting product. In this case the titration curve would be displaced to the right. From the linear extrapolation to the volume axis, the value of a_t can be used to determine the amount of material stoichiometrically bound to the measured ion, even though this material itself is not indicated by any ion-selective electrode.

In the first case (ion to be determined is indicated and added) we are dealing with a *successive addition* method. Because of the large number of data points and the graphical method of analysis a *much greater accuracy* can be attained than with the single standard addition methods.

The second case (ion to be determined is not directly indicated, but reacts quantitatively with an indicated titrant ion) is a regular titration, in which only the data in the region *after* the end point, where the indicated ion is present in sufficient concentration, is employed in the evaluation. This procedure of working with an *excess of titrating agent* is of special advantage for precipitation titrations, since an excess of precipitating agent is present in solution. For a 1:1 reaction in which

$$a_{\text{ion to be determined}} \cdot a_{\text{precipitating agent}} = K_S \tag{90}$$

this means that the equilibrium activity of the ion to be determined is smaller. Hence even when working with smaller ion concentrations, precipitation is more likely to occur using this technique. This is of special advantage for the determination of low concentrations, since otherwise ion concentrations of lower than $\sqrt{K_S}$ would not lead to precipitation. An additional problem can sometimes arise in precipitation titrations: *The precipitate solubility may change with time.* An example is LaF_3, whose solubility product changes from 10^{-15} to $10^{-17.6}$. This process completely obscures any normal titration curve, as shown in Fig. 50, rendering a concentration evaluation impossible [233]. Only with the linearized plot of Gran's shown below is an evaluation possible. This graphically illustrates the advantages of this method.

The same procedure (determining an ion not directly indicated through quantitative reaction with an indicated ion) can be extended to an indirect titration by adding a known amount of the non-indicated ion to an excess of a second, non-indicated material with which it reacts quantitatively, and then back-titrating this excess with some directly indicated ion.

Phosphate, for example, can be determined in this way by precipitation with excess lanthanum, and then back-titrating the remaining lanthanum with fluoride ion as usual.

In instances where the ion to be determined is directly indicated, one of two techniques may be chosen: The addition technique described above can be used, or a titration can be carried out in which the amount of indicated ion is quantitatively diminished. In this case, of course, the Gran titration curve will increase from right to left. For trace analyses the first technique should be used, and with larger amounts of material the second is more convenient. With very high concentrations, large end point jumps and negligible symmetry errors, the usual titration method is recommended for the simpler evaluation procedure.

Certain special precautions must be observed when analyzing complexometric titrations with Gran's extrapolation method. The complex formation constant is dependent on the pH value, since hydrogen ions compete with the titrated metal ion for ligand binding sites. In highly acidic solutions effectively all of the ligand sites are occupied by H^+ ions due to their excess. In this case metal ion complexes cannot be formed to any great extent at metal ion concentrations ≤ 0.1 M. In too highly alkaline solution the danger exists that the metal ion may precipitate as the hydroxide. Hence, one works in an intermediate pH range (more details in the special literature [234]) with complexometric titrations. In such a pH range some ligand sites (in poly-

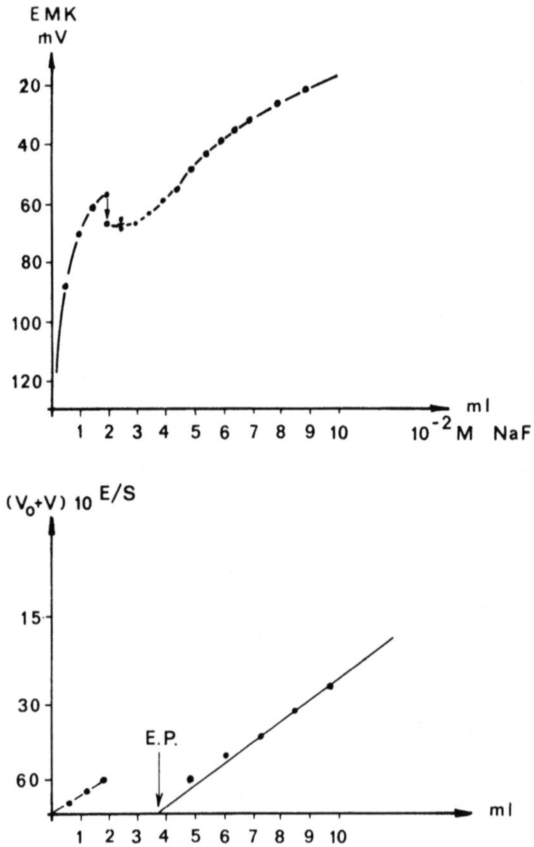

Fig. 50. Titration of traces of La^{3+} with NaF. Above: Normal titration curve; Below: Extrapolation according to Gran

dentate ligands such as EDTA) are partially occupied with hydrogen ions. Only when sufficient energy is released upon complex formation can these hydrogen ions be displaced:

$$Me^{2+} + EDTA\text{-}H_2 \rightleftharpoons Me\text{-}EDTA + 2\,H^+. \qquad \text{(R38)}$$

To shift this reaction as far to the right as possible, the hydrogen ions produced must be removed from the equilibrium (law of mass action). This can be done with a suitable pH buffer system (acetate, hydrogen phosphate, citrate, etc.). Unfortunately, these added buffer systems sometimes also form weak complexes with the metal ion to be measured. This can easily be confirmed with the corresponding metal cation-indicating ion-selective electrode. In this case the electrode potential in the pH buffer system-metal ion solution will be more negative than that in a pure metal ion solution of the same concentration.

As an example, in an acetate buffer, cadmium exists only 10% as the free ion; in a citrate buffer only 0.01% [234]. Normally this does not hinder a complexometric titration because the titrant usually forms a substantially more stable complex with the measured ion. Therefore even the weakly complexed fraction of the ions are usually indicated. With the Gran evaluation procedure, however, a special property of metal ion-pH buffer complexes must be noted: *These metal ion-pH buffer complexes function as buffers for the corresponding metal ion.* As shown in Sect. 5.4.1, in the case of complex formation the activity of the excess free metal ion is proportional to the ratio of total metal ion concentration to total ligand concentration. With a large excess of pH buffer system, a volume change of about 10 to 15%, roughly that resulting from titrant addition, causes practically no change in this ratio. *An ion-selective electrode will show no effect of diluting the solution in this case.* For this reason the volume correction can be dispensed with when using the Gran analysis procedure with complexometric titrations in excess pH buffer. In practice, this means that if complex formation between metal ion and pH buffer has been confirmed through a negative shift in electrode potential, one must be sure to provide an excess of this buffer.

It is also to be noted that polydentate chelating agents can sometimes form relatively strong complexes of different stoichiometry at the beginning of a titration, when the metal ion is present in large excess. Under such circumstances the Gran function $10^{E/S}$ will have a different slope in this region than it will in the region shortly before the end point. The stoichiometry of the complexes formed in this case can be determined from the extrapolation of the different Gran lines [235]. In this way one can obtain deeper insights into the fundamentals of the corresponding complex formations.

Summarizing once again the merits of the Gran technique:

From a technical standpoint, this method has the advantage that the EMF corresponding to the equivalence point need not be directly measured via a titration. We have seen that with the small measured ion concentrations in this region, large uncertainties (also long response times) in the EMF measurement are possible (indication of interfering ions, since the $\sum_{\mathrm{I}} k_{\mathrm{M-I}}^{\mathrm{pot}} \cdot a_{\mathrm{I}}$ term in the empirical Nernst equation is too large compared to the measured ion activity). The same holds for direct potentiometric measurements near the detection limit. While waiting for a stable potential reading, a small amount of measured ion dissolves out of the active phase, falsifying the final result. In the addition method of Gran, the dissolved amount of measured ion can be ascertained with a blank solution by extrapolating back to the otherwise elusive zero value, and can subsequently be corrected for.

This method of determining the titration end point is much more reliable due to the extrapolation procedure. The b value (see Sect. 5.3.2) is also far less influential in such titrations. In the Gran analysis a large b value does not give rise to systematic errors, but rather to a poorer reproducibility if it is much greater than 1. Asymmetric titration curves are also hardly a problem, since in this method the evaluation does not place great importance on the inflection point arising between two different mathematical functions intersecting at the equivalence point, but rather extrapolates to the end point from only one side of the titration curve.

In addition to overcoming all these problems in the actual measurement and in the evaluation method, this technique also avoids a series of chemical titration problems. These arise with other techniques when too large a solubility product or too small a complex formation constant in dilute solutions result in poorly defined end points in the usual titration curves. With the Gran method, with which an excess of titrating agent can be used, these phenomena only cause a deviation from the titration line immediately before the equivalence point. This, however, in no way hinders an accurate extrapolation. We have already seen that even a time-dependent precipitate solubility product does not substantially influence the evaluation in precipitation titrations. Even the determination of different halide ions in the same solution, which is known to lead to co-precipitation, can be accomplished more accurately with the Gran method. First, an early portion of the calibration curve is employed, where the precipitation is not complete, and thus the co-precipitation is not yet maximal. Next, the less soluble precipitate can be ascertained by a further addition process. The total halide content, which is correct in all cases, can be attained by determining the silver ion content after passing the chloride equivalence point.

It is more difficult to point out the disadvantages of this technique. The fact that only about one order of magnitude concentration change is spanned when using this method is due simply to a drawback of the linear plotting scale. For each further order of magnitude in concentration, a ten-fold longer abscissa scale is required. *Although in principle a line can be constructed with only two points, the statistical error of the extrapolation is reduced with each additional data point.* In plotting the complete titration course with this method, in addition to the deviation shortly before the end point, one sometimes finds another just at the beginning of the titration, which implies a certain supersaturation of the solution. The question then naturally arises as to which portion of the titration curve should be chosen for the evaluation, since with beginning and end curvatures the job of constructing a straight line is basically one of drawing a tangent. *It is for precisely this reason that, in the interest of better statistical error limits, many more data points than just three or four be obtained. With a larger number of data points one is in a much better position to be able to determine where the curvatures begin to set in, and thus to choose the portion of the curve to be used for the evaluation.*

Since this procedure practically allows the display of only a single power of ten concentration range, the mV range within which all these data points must lie is only about 60 mV. In order to obtain a large number of *significant* data points in such a small mV range, a cell should be constructed with which an accuracy of at least 0.1 mV can be attained. A cell with any greater precision is of no avail for the reasons discussed in Chap. 4; this has also been borne out experimentally [236].

In using the commercially available graph paper for carrying out Gran method analyses, one must keep in mind the fact that the ordinate calibration assumes a Nernst slope of 58 mV/power of ten. This value corresponds to that most often found in practice. Nonetheless, for accurate analyses one should ascertain the effective slope of the indicating electrode by constructing the usual calibration curve. The error arising from a change in slope of the titration curve due to a false electrode slope can indeed be calibrated for through a standard solution titration, but in doing so one surrenders the advantage of having eliminated systematic errors with the Gran method.

Concerning the graphical analysis, be aware that a read-off error of as much as 0.5% must be reckoned with, depending on the size of the graph and the thickness of the line. An advantage of the graphical method is the automatic averaging and the identification of stray points. With increasing use of data processing equipment, especially in analytical laboratories, the Gran method may find wider application. If a computer is programmed to consider only the region between 30 and 80% of the full titration curve and to ignore about 10% stray points (for their origin see Chap. 4), then under ideal conditions with an EMF accuracy of 0.1 mV, in the titration of a 0.01 M solution, an accuracy of 0.1%, and for a 10^{-3} M solution titration an accuracy of 0.5% can be reproducibly attained with this method [236].

Under practical circumstances (with interfering ions and chemical complications) an error of up to 3% must be dealt with here as well [237]. In this connection a comparison of the different methods, all applied to the same titration (in this case a simple titration of 10^{-2} M chloride solution with 0.1 M $AgNO_3$) may prove instructive. Isbell et al. [238] developed a special computer program, "TITRATE" (IBM 360/65), to evaluate the end point of this titration and compared its accuracy with those obtained using the method of tangent intersection, the first and second derivative methods and the Gran plotting procedure. They came to the following conclusions:

TITRATE	error ± 0.1%
tangent intersection	error + 1.5%
first derivative	error + 5.4%
second derivative	error + 5.4%
Gran method	error − 2.4%

5.8 Practical Example of the Gran Extrapolation Method: Determination of Chloride in the ppm Range

As a representative of the large number of possible determinations which can be carried out with this technique, one will be described here in detail. Since the basic operations are the same, this procedure can be easily extended to other ions and titrating solutions.

Determination of the smallest traces of chloride has become increasingly important in materials testing and environmental protection. Iron and steel corrode more easily when chloride ion is present. For two industrial steel applications, the exclusion of chloride ions is extremely important for this reason. One is the field of energy conversion by means of high pressure steam turbines. Here the chloride content of the boiler feed water must stay below 1 ppm in order to avoid premature corrosion. The second field is "drooping" of reinforced concrete constructions. Here, above all, the HCl librated from burning PVC can infiltrate the pores of the concrete and find its way to the structural steel reinforcements. The resulting corrosion represents a serious safety hazard. Since PVC is becoming increasingly employed in electrical insulation, after every major fire it is necessary to test the extent to which other parts of the building not directly on fire have been poisoned with chloride.

Chloride content is also of great importance in clinical diagnoses. Here a sensitive procedure means that less sample volume is required, an especially desirable quality when working with infants and small children.

Required Apparatus:

1 pH or ion meter with a precision of at least 0.1 mV
1 silver ion-indicating Ag_2S solid-state membrane electrode
1 Ag/AgCl reference electrode with sleeve diaphragm and double salt bridge
1 thermostated (< 10° C) sample cell
For a titration with added $AgNO_3$ solution:
1 precisely calibrated 1 ml micrometer burette
For a coulometric titration:
1 constant current source, 0 to 100 mA
1 silver wire (≥ 99.9%)

5.8.1 Principles

For reasons of accuracy we choose a titration procedure using silver ion as titrating agent. We know that the precipitate AgCl resulting from the titration has a solubility product of $K_S = 10^{-9.75}$ in water at 25° C, a relatively large value. We can easily calculate the amount of chloride remaining in solution:

$$a_{Cl} = \frac{K_S}{a_{Ag}} ; \tag{91}$$

at the equivalence point $a_{Cl} = a_{Ag}$, so with the K_S reported above, a chloride ion activity of $10^{-4.9}$ M results. This corresponds to 0.35 ppm chloride, so amounts under 0.35 ppm will not lead to precipitation, and thus cannot be titrated, We have seen that by using the Gran method, this detection limit can be decreased by a factor of ten with a ten-fold excess of silver ion according to Eq. (91). Since we are interested in precise chloride determinations in the sub-milligram region, we would like to chemically lower this solubility product as well. We know that less polar solvents than water are poorer dissolving media for ionically bound compounds. It has been shown that the solubility product of AgCl in 54.2 wt. % methanol is $K_S = 10^{-11.38}$, in 54.15 wt. % acetone $K_S = 10^{-12.13}$ and in 60.73 wt. % dioxane $K_S = 10^{-12.36}$ [239]. This lowering of the solubility product by almost three orders of magnitude corresponds to an equilibrium equivalence point chloride content of less than 0.035 ppm. Similar results were found using 100% acetic acid as the solvent [240]. Considering the 9:1 dilution of an aqueous sample with 100% acetic acid to achieve a 90% acid concentration in the test solution, a decrease in the lower detection limit of greater than a factor of ten relative to that in pure water can be attained. We choose 90% acetic acid as solvent here, since this concentration also exhibits a sufficient conductivity. For the same reasons considered above we will thermostat our solution at under 10° C.

5.8.2 Preparatory Work

It need not be pointed out that in this example, as for all trace analyses, extremely pure chemicals must be used. Acetic acid is available in extremely pure form (Puranal, Suprapur). In chloride trace analysis special care must be given to the electrolyte contact between reference element and sample solution, since most reference half cells are filled with chloride-containing solutions. We will use a 0.1 M KCl solution as primary electrolyte for our Ag/AgCl reference electrode, and a KNO_3 saturated 90% acetic acid solution for an outer electrolyte, which is to be renewed daily. In this way we have also made all provisions necessary for a constant liquid junction potential (see Sect. 2.3);

We wish to start out with 10 ml of sample solution and dilute it with 90 ml of 100% acetic acid. We will also want to adjust the concentration of the $AgNO_3$ solution, so that a sample solution chloride content of 1 ppm requires exactly one burette full of this solution (= 1.000 ml). This requires a 2.82×10^{-4} M silver nitrate solution. Due to the difficulties in storing such dilute silver ion solutions, freshly prepared solutions should be employed, or preferably the more elegant method of coulometric reagent production (details in [241]) should be used.

Following these fundamental considerations, we determine the slope of the Ag_2S solid-state membrane electrode used in 90% acetic acid, since we will need this value for plotting the Gran function $10^{E/S}$. The concentration range 10^{-2} to 10^{-7} M $AgNO_3$ is sufficient. Finally, we measure the EMF of a cell containing a suspension of a spatula tip full of AgCl in 90% acetic acid. Through extrapolation of the calibration curve thus obtained, we can ascertain the Ag^+ ion content in solution due to the equilibrium with solid AgCl, and thus determine its solubility product in this medium. Naturally, an equivalent amount of Cl^- ions are also present in solution. This amount represents the smallest chloride ion concentration which must be surpassed in order for precipitation to occur. In order to titrate even smaller quantities of chloride, one must start with such a suspension of AgCl. The procedure has the additional advantage that the suspension provides many nucleation centers, which helps to avoid supersaturation during the course of the titration.

5.8.3 Blank Determination

10 ml distilled water are placed in a 150 ml beaker with a magnetic stirrer, and 90 ml 100% acetic acid (extremely pure) are added. About 10 mg of AgCl are scattered over the solution, and the titration with 2.82×10^{-4} M $AgNO_3$ solution is begun with stirring. After the addition of each 0.025 ml the electrode potential is recorded. For the very low concentrations of measured ion (about 10^{-8} M Ag^+ ions) somewhat longer response times (about 1 minute) are required. After 0.5 ml of titrant have been added (equivalent to a chloride content of 0.5 ppm), the titration is stopped and the EMF values thus far obtained are used to construct a plot of the Gran function (ordinate, $10^{E/S}$) against the added titrant volume (linear abscissa). With this procedure the dilution effect during the course of the titration (100 → 101 ml) can be neglected. If no impurities are present in the reagents used, then the Gran function passes through the origin. If the Gran function is displaced to the right, then this amount (1 ml ≙ 1 ppm) corresponds to the background chloride level of the solutions and apparatus used. In this case the analysis should be repeated a number of times to insure that this value is reproducible, because only then is a correction of the chloride determinations to follow possible. To check the titer (1 ml ≙ 1 ppm) of the silver nitrate solution, a freshly prepared standard solution with 0.5 ppm chloride content should be titrated.

Due to the strong adsorption properties of glass, plastic beakers should be employed throughout.

5.8.4 Chloride Content Determination of the Sample Solution

Instead of the 10 ml of water in the blank determination described above, the sample solution to be investigated is used here. Otherwise, the procedure is exactly the same as before. Silver nitrate solution is added to the sample solution until the same potential as that found in the blank determination is attained. Here also the Gran function is plotted to an excess of about 0.5 ml silver nitrate solution. The chloride content of the solution can then be read directly off the graph from the point where the curve intersects the volume axis (in ppm for a titer of 1.00 ppm/ml). The previously determined blank value is then subtracted from this number.

5.9 Determination of Some Characteristic Electrode Parameters

Next to the detection limit, the most important piece of information for the potential ion-selective electrode user is the selectivity. Because of the large number of possible interfering ion combinations, the manufacturers only report the selectivity coefficients for a few interfering ions. As we will see, the selectivity coefficients depend both on the total ionic strength and on the particular method of determination. Unfortunately, the method used to obtain these selectivity coefficient values is not always indicated. For these reasons, the reported values can only serve as approximate indications of the selectivities to be observed under a given set of experimental conditions.

In the interest of a satisfied clientele, trustworthy manufacturers allow their electrodes to be tested for suitability before purchase. For this reason the following sections describe procedures for determining the most important electrode parameters, so that the prospective customer can test to see if a specific electrode is suitable for his particular problem before making a purchase.

5.9.1 Determination of the Detection Limit

With the proper sample preparation (see Chap. 5.1.2), an ion-selective electrochemical cell can, as a rule, span many orders of magnitude in concentration before a large deviation from the calibration curve (EMF vs. log c) arises.

Following the suggestion of the IUPAC [455], the detection limits with ion-selective electrode techniques are defined analogously to other signal-producing physical chemical techniques as that measured ion concentration, at which the measured signal is exactly twice as large as the background noise. This is precisely the case when the deviation from the Nernst line is $18/z$ mV ($59/z \log 2 = 18/z$ at 25°C) (see Fig. 51)[1]. The

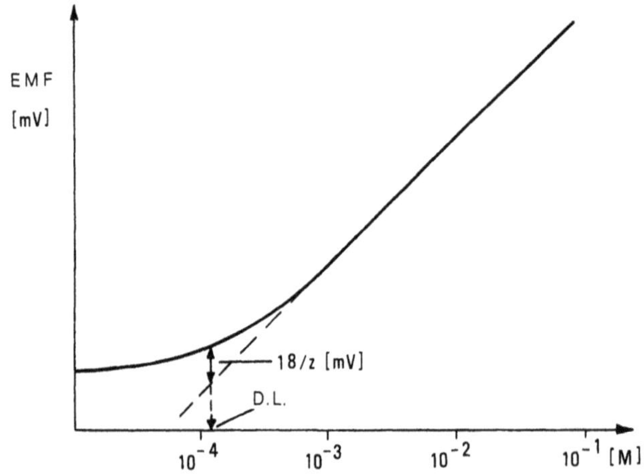

Fig. 51. Determination of the detection limit

[1] Liteanu et al. [456] however showed that a more thorough statistical approach leads to actual detection limits of over one order of magnitude smaller (s. also Chapt. 5.5.2).

detection limit of an ion-selective electrode is influenced by many factors. Theoretically, the detection limit is determined by the solubility of the electroactive electrode phase, which contains some relatively firmly bound measured ion, in the sample solution. No smaller ionic activity than that corresponding to this solubility product can be measured. In this context it should be noted that complexing agents can lead to poorer detection limits (Example: EDTA can react with the LaF_3 crystal of the fluoride electrode to complex La^{3+}, setting the corresponding amount of F^- free).

In general, this theoretical detection limit is not achieved, either because such extremely dilute solutions are unstable due to irreversible adsorption on the container walls, or in fewer cases another ion becomes potential-determining.

When working with conditioning solutions, impurities in the chemicals used to prepare these solutions can determine the detection limits.

The detection limit for a particular application of interest is easy to determine. One starts with solution free of measured ion, whose matrix resembles that of the sample solutions to be tested, and adds successive portions of measured ion to span the range 10^{-7} to 10^{-1} M. A calibration curve is constructed from the data thus obtained, and the detection limit is evaluated as shown in Fig. 51. If conditioning solutions are later to be used (to free bound measured ions, bind interfering ions, etc.), they should also be used here, of course. If, as is to be expected, the EMF value in the measured ion-free solution drifts, the value after three minutes should be recorded.

The determination of the actual detection limit will show that this limit is made poorer by the presence of interfering ions. The observed detection limit can be improved if one exactly calibrates for the lower curvature of the calibration curve.

5.9.2 Determination of the Selectivity Coefficient

As is the case with any analytical technique, interferences can arise when working with ion-selective electrodes. These interferences can generally be classified as interferences due to direct indication of other ions or interferences due to chemical influences on the measured ion, such as complexation, which only alter the activity of *free* measured ion indicated by the electrode. This second type of interference is not due to the electrode itself, but rather to the solution chemistry of the corresponding measured ion in the particular matrix employed, and as such will not be considered in any further detail here.

The indication of an interfering ion can be quantified by means of a selectivity coefficient. This coefficient represents the sensitivity ratio: Interfering ion/measured ion. Thus, for a selectivity coefficient $k_{M-I}^{pot} = 10^{-3}$, the measured ion is indicated 1,000 times more sensitively than the interfering ion; i.e. a 1,000-fold higher activity of interfering ion is needed to produce the same EMF as a given activity of measured ion. This interfering ion indication can be mathematically described (approximately) with the Nernst equation (22) described in Chap. 1.5. There are a number of possibilities for experimentally determining these selectivity coefficients.

Constructing a calibration curve for any particular interfering ion usually results in a picture qualitatively the same as that shown in Fig. 52. With such a graph there are two methods for determining the selectivity coefficient.

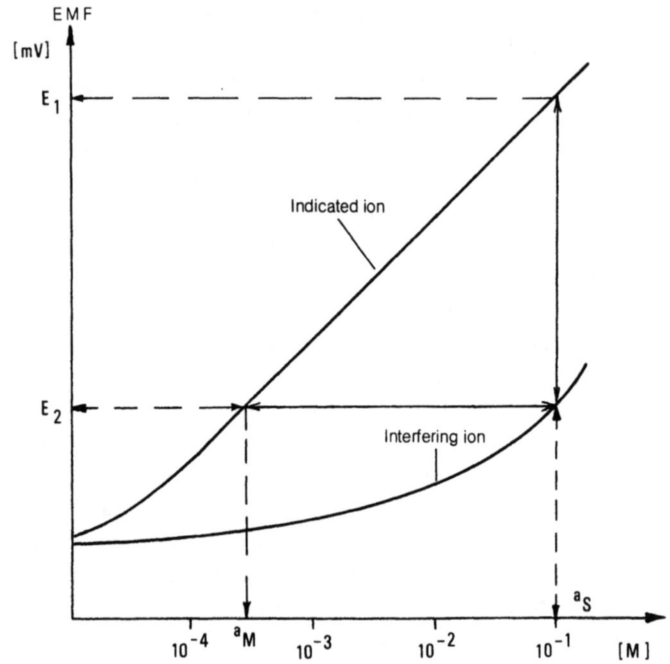

Fig. 52. Determination of the selectivity coefficients according to the method of separately measured solutions

At the same measured and interfering ion activities k^{pot}_{M-I} can be calculated from the following equation:

$$\log k^{pot}_{M-I} = \frac{(E_2 - E_1)z \cdot F}{2.3\, R \cdot T} = \frac{\Delta E}{S} \tag{92}$$

where S = slope, which should be identical for both ions.

k^{pot}_{M-I} can also be calculated from the measured and interfering ion activities which give rise to the same cell EMF according to:

$$k^{pot}_{M-I} = \frac{a_M}{a_I}. \tag{93}$$

It is easy to see from Fig. 52 that the selectivity coefficient itself is a function of the absolute measured and interfering ion concentrations. The first procedure is suitable for quick rough measurements since only two measurements and two solutions are required. Both procedures are unrealistic in that in practice, one is dealing with a mixture of measured and interfering ions. The IUPAC [455] has recommended a method in which a constant interfering ion activity and a variable measured ion activity are employed. The measured EMF values are used to construct a calibration curve as usual (Fig. 53). To evaluate k^{pot}_{M-I} one first determines the intersection point of the two

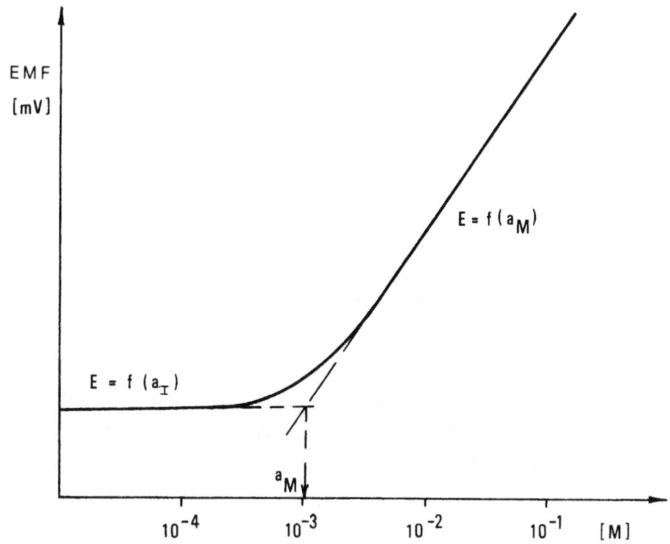

Fig. 53. Determination of the selectivity coefficients according to the method of mixed solutions

linear portions of the curve [i.e. the point at which both terms in the logarithm of Nernst Equation (22) are equal]. From the measured ion activity a_M read off of the activity axis and the constant interfering ion activity a_I, the selectivity coefficient can be calculated according to Eq. (93). Here again, the selectivity coefficients are concentration dependent, so the activity of the interfering ion must be specified whenever reporting a k_{M-I}^{pot} value.

In addition the selectivity coefficient depends on the method of determination, therefore this information should also be included. Because some electrodes show a considerable EMF drift and different slopes when exposed to pure interfering ion solutions, the IUPAC method is strongly recommended.

With respect to the comparability of selectivity coefficients obtained by the various methods, Liteanu et al. [457] recently proved with the aid of a strictly statistical treatment that in general the separate solution methods yield "meaningless" results!

The dependency of the potentiometric selectivity coefficient on the method of determination can also be explained by the mixed potential theory [448]. In the separate solution method the thermodynamical equilibrium can be achieved for both ions (if the interfering ions show a similiar slope it is a good indication for this). In such a case one will find a good correlation of the k_{M-I}^{pot}-values with other thermodynamic parameters, e.g. stability constants and distribution coefficients. In the case of the mixed solution method however, the different electrode reactions, which should lead to the thermodynamic equilibrium, all start at the same time (parallel reactions). Therefore the fastest reaction, which is the one with the highest exchange current density, will mainly determine the interfacial potential and is likely to reach the thermodynamic equilibrium potential first. This potential then acts on the other electrode reaction like an externally applied overvoltage disturbing the thermodynamic equilibrium and thus leads only to a steady state situation for the interfering ion.

6 Applications of Ion-Selective Electrodes

The examples listed in Table 8 on the following pages are in no way claimed to be a complete survey. The fact that biochemical applications are given somewhat more weight here is because, in this author's opinion, it is in this area that the advantages of the ion-selective electrode analysis technique (direct activity determination with miniaturized sensors) are most convincingly demonstrated, especially in view of the complexity of the sample media. For the analytically trained laboratory chemist, the capabilities and limits already discussed in the text should suffice to allow him to develop his own new applications. In industrial applications ion-selective electrodes are being employed with increasing frequency. With the proper "know-how", these ion-selective electrodes are not nearly as unsuitable as a few early bad experiences would lead one to expect.

As pointed out in the Preface, it is not intended to provide any "cookbook recipes" here. The activity in this field is so enormous today, that by the time such procedures appeared in print they would already be outdated. In addition, the more trustworthy manufacturers often include the best, most up-to-date working procedures with the electrodes themselves.

6.1 Physiology, Biology, Medicine

As mentioned in the Introduction, the fields named above offer a wide variety of possible applications. These have not yet been taken full advantage of. Ever since the classical investigations of McLean and Hastings in the years 1934 and 1935 [242, 243], which established a direct connection between the contraction amplitude of a beating, isolated frog heart and the Ca^{2+} ion activity of the solution used, it has been recognized that the activity rather than the concentration is the determining factor in many physiological processes. Thus, in order to investigate the causal factors in such systems, the *activities* of the participating ions must be known. Most analytical procedures, however, supply the concentration as a final result. In order to arrive at the corresponding activities, the individual activity coefficients are needed as correction factors (see Appendix). In contrast to dilute, aqueous electrolytes, in which the mean activity coefficients of cation and anion can at least be estimated with the Debye-Hückel approximation, there is still confusion as to the mean activity coefficients of ions in biological fluids containing protein molecules, polyelectrolytes and colloidal materials. To aid in elucidating such problems, experimental probes must be employed which respond directly to the ionic activity (osmotic pressure, potentiometry).

Table 8. Examples of ion-selective electrode applications

Species/Sample matrix	Sample preparation	Electrode	Analysis technique	Literature
Aluminum				
general	0.2 M $NaNO_3$, NH_4NO_3 or 0.01 M $Mg(NO_3)_2$+CuDCTE	Copper	B/DCTE	[260]
geology	dissolve gibbsite, Hydroxide, sulfate	Fluoride	D	[261]
metallurgy	dissolve metal or alloy in HCl, acetate buffer pH 3.8–4.6	Fluoride	T/NaF	[224, 262]
radiochemistry	pH 3.8, 1:20 dilution with ethanol	Fluoride	T/NaF	[263]
Ammonia				
plasma	1 + 4 dilution with dil. NaCl, pH 11.3 with dil. NaOH	NH_3	D	[264]
sea water	pH > 11	NH_3	D	[434]
soil	ground, air-dried soil suspended in 2 M KCl, filtered, pH > 12 with NaOH	NH_3	D	[265]
urine	dilute with dist. water, pH > 12 with NaOH	NH_3	D	[264]
waste gas	adsorb in dil. H_2SO_4, then pH > 12 with NaOH	NH_3	D	[264]
waste water	filter sample, dilute 1 + 9 with 0.1 M NaOH	NH_3	D	[264]
Ammonium				
general	convert to NH_3 at pH > 12 with NaOH	NH_3	D	[264]
	pH ~ 7, buffer	Cation	D, T	[266, 399]
liq. NH_3		Cation	D	[267]
serum	50–150 μl samples, pH > 12 with NaOH	NH_3	D	[167]
Arsenic	convert to As^{5+}, add excess buffered $La(NO_3)_3$ solution, dilute with water and back titrate excess La^{3+} with NaF	Fluoride	T/NaF	[264]
Azide		Silver	$T/AgNO_3$	[268]
Barium	treat sample solution with excess Na_2SO_4, back titrate with $Pb(ClO_4)_2$	Lead	$T/Pb(ClO_4)_2$	[264]
	add CuDCTE	Copper	B/DCTE	[264]
	direct	Barium	D	[136, 157]
Beryllium		Divalent	D	[264]
Boron				
glass	dissolve powdered enamel in HF, pass through ion-exchanger	BF_4^-	D	[264]
plants	convert to BF_4^-	BF_4^-	D	[264]
soil	extract soil suspension, convert to fluoride form with ion-exchanger	BF_4^-	D	[273, 274]

Table 8 (continued)

Bromide				
atmos. ppt.	pH 2–12	Bromide	D	[276]
Photoemulsions	pH 3–4 with HNO_3	Bromide	D	[264]
nonaq. solvents	alcohol and acetic acid	Bromide	D	[277]
vegetable oil	extract with ether and liberate the bromine	Bromide	D	[278]
Bromine				
aq. solution	add known amount of iodide and measure its decrease	Iodide	S	[264]
organic	burn sample at 1000°C and absorb gases in a HNO_3, acetic acid, H_2O_2 solution	Ag_2S	T/Hg^{2+}	[275]
Cadmium				
galvanic	Cyanide complex; treat with excess hypochlorite solution, pH 10	Cadmium	S	[272]
general	pH 3–12, titrate with 1,10-phenanthroline	Copper/CdS single xstal	T	[270, 279]
Calcium				
beer	filter, pH 5.6–6.0 with NaOH and methyl orange	Calcium	S	[264]
blood and other biological fluids		Calcium	D	[250, 312, 315–328]
boilers	dilute 1:1 with NH_4OH, add CdEDTA indicator	Cadmium	T/EDTA	[264]
grain	ash at 600–700°C, dissolve in HCl, establish pH 10, add CdEDTA indicator	Cadmium	T/EDTA	[264]
milk	periodic cleaning of the electrode with Na-dodecylsulfate + EDTA solution, otherwise no preparation	Calcium	D	[329, 330]
minerals	dissolve in HNO_3, pH 5.5 with NaOH, add excess EDTA;	Cadmium	T/Cd^{2+}	[264]
	in the presence of Mg add CuEGTA	Copper	T/EGTE	[267]
sea water		Calcium	D	[281, 282]
serum	add 5–10 units heparin	Calcium	D	[283–314, 458]
soil	extract dry soil with sodium acetate solution at pH 8.2, centrifuge	Calcium	D	[280]
sugar solution		Calcium	D/S	[334]
water	buffer at pH 10	Calcium	T/EDTA	[331, 332, 333]
wine	dilute 1 + 9	Calcium	S	[264]

Carbon	total inorganic + organic C in water	CO_2	S	[431]
Carbonate	measure CO_2 set free upon acidification with $HClO_4$	CO_2	D/S	[167]
	pH 7.5 and measure directly with special liquid membrane electrode	CO_3	D	[146]
Cesium	pH 7	Potassium/ cation	D	[264]
Chloride				
air	decomposition product of PVC	Chloride	DB	[340, 276]
boilers	add 5 M HNO_3	Chloride	S	[335]
	in acetic acid: titrate with $AgNO_3$	Ag_2S	T/Ag^+	[s. page 155]
cheese	blend 2 g with 50 ml water	Chloride	D	[339]
galvanic		Chloride	$S,T/Ag^+$	[272]
geology	decompose minerals with Na_2CO_3/ZnO, acidify with HNO_3	Chloride	D	[336]
	dry soil samples: leach with H_2O; filter	Chloride	$D,T/Ag^+$	[337, 338]
milk		Chloride	D	[341]
perspiration	direct measurement on skin surface in the diagnosis of cystic fibrosis	Chloride	D	[347, 348, 349]
plants	shake sample in HNO_3	Chloride	T/Ag^+	[342]
preserved foods	add 0.5 ml formaldehyde per 100 ml sample, mix, dilute 1:20, add HNO_3 1 + 4	Ag_2S	T/Ag^+	[264]
sea water	use calibration solutions made up in sea water matrix	Chloride	D	[343, 344, 345]
serum	measure directly with flow measurement	Chloride	D	[264]
urine	dilute urine 1 + 3 with acetic acid	Ag_2S	T/Ag^+	[346]
Chlorine				
active	add excess of iodide solution and measure decrease in iodide concentration	Iodide	S	[264]
organic	burn sample at 1000°C and absorb gases in HNO_3, acetic acid-H_2O_2 solution	Ag_2S	T/Hg^{2+}	[275]
Chromium	convert to chromate, add excess iodide and measure the decrease in its concentration	Iodide	S	[264]
galvanic	dilute chromate solution 1 + 99 with 1 M acetic acid	ClO_4^-	D	[272]
Citrate	establish pH 10–12 with NaOH	Copper	T/Cu^{2+}	[264]

Table 8 (continued)

Cobalt	bring pH of Co^{2+} solution to 10 with NH_4OH, add CuEDTA	Copper	T/EDTA	[264]
Complexing Agents		Copper	T/$Cu(NO_3)_2$	[264]
Copper				
galvanic	mask Fe^{3+} with NaF	Copper	D	[272]
Cyanate	heat 30 min. in 50% H_2SO_4 to hydrolyse to NH_3, pH > 12 with NaOH	NH_3	S	[264]
Cyanide				
cigarette smoke	adsorb smoke on an Ascarit-Filter, extract with NaOH, add $Pb(NO_3)_2$	Cyanide	D	[354]
galvanic	remove strongly complexing ions with an ion-exchanger, dilute with 0.1 M NaOH	Cyanide	D	[272]
plants	mix sample with Sorensen phosphate buffer, add β-glucosidase, heat, add NaOH	Cyanide	D	[353]
waste water	direct monitoring	Ag_2S/CN^-	D	[350, 351, 352]
Fluoride				
acids	dilute with conditioning solution, pH 5.5	Fluoride	D	[369]
air	collect on prepared filter, extract with H_2O, add conditioning solution	Fluoride	D	[357–360]
foodstuffs	ash, dissolve and add conditioning solution	Fluoride	D,S	[367, 368]
galvanic	dilute with conditioning solution	Fluoride	D	[355, 356]
geology	fuse 1 g mineral, rock or soil sample with 10 g Na,KCO_3, dissolve in HCl, dilute to 250 ml, dilute again with special conditioning solution 1 + 9 (citrate buffer pH 6 and high Titriplex-IV concentration) and measure directly	Fluoride	S	[361]
plants	decompose with Na_2CO_3/ZnO, dissolve in water, dilute with conditioning solution	Fluoride	D,S	[359, 366]
sea water	dilute with conditioning solution	Fluoride	D,S	[362–365]
serum, biol. liquids	inorg. fluoride: simply dilute with conditioning solution	Fluoride	D	[370–375]
	total fluoride: ash with $Na_2CO_3/MgCl_2$, separate fluoride via diffusion	Fluoride	D	
toothpaste	dissolve sample in $HClO_4$, add conditioning solution	Fluoride	D,S	[376–380]
waste water	add pH 5.5 buffer	Fluoride	D	[264]
Gold	establish pH 1.2–2.3	Iodide	S	[381]
galvanic	dilute with NaOH and titrate with $AgNO_3$; shows two end point jumps for both free and bound CN^-; total-free CN^-=Au	Ag_2S	T/Ag^+	[272]

Iodide				
foodstuffs	pulverize in hot water, cool, add 10% phosphate solution to establish ionic strength	Iodide	D	[383]
general	90 vol% acetone	Sodium	T/Ag^+	[382]
hydrogen peroxide	no preparation	Iodide	D	[388]
pharmacy	establish pH 9–11 with NaOH, reduce iodate to iodide with Al	Iodide	D	[384, 385]
radioactive ^{131}I		Iodide	D	[386]
selenium	decompose with NaOH, neutralize with H_2SO_4, filter	Iodide	S	[387]
Iron	titrate Fe^{2+} with CN^-, Fe^{3+} with F^-	CN^-, F^-	T/CN^-, F^-	[224]
Lanthanum	neutral, unbuffered solution, 60% ethanol	Fluoride	T/NaF	[389, 390]
Lead				
biology		Lead	D,T	[271]
galvanic	dilute 1 + 1 with methanol	Lead	T/Na_2MoO_4 or Na_2WO_4	[272]
general	pH 4–7	Lead	D	[269]
	pH 4–7	Copper	T	[270]
paint	dissolve in HNO_3, buffer pH 5–6, dilute 1 + 1 with methanol	Lead	S	[264]
Lithium	neutralize with NH_4OH if neccessary, precipitate interfering ions with $(NH_4)_2S$, dilute with alcohol	Fluoride	T/NH_4F	[391]
Magnesium				
biochemistry	bound to nucleic acids: neutralize with 5 mM sodium phosphate buffer	Divalent	D	[392]
sea water	calibrate with "synthetic" sea water	Divalent	D,S	[393, 394]
sugar beets	establish pH 11, titrate with EDTA and calcium electrode to Ca endpoint, then further with Divalent electrode to Ca + Mg endpoint	Calcium/ Divalent	T/EDTA	[345]
Manganese	treat Mn^{2+} solution with CuEDTA solution	Copper	T/EDTA	[264]
Mercaptans				
water soluble	dissolve in water or dilute HNO_3	Ag_2S	T/Ag^+	[395]
water insoluble	dilute 1 + 9 with 75% methanol or acetic acid	Ag_2S	T/Ag^+(methanolic)	[264]

Table 8 (continued)

Mercury	titration with bromide	Bromide	T/Br^-	[224]
organomercurials	destroy with acid, neutralize	Iodide	D	[264]
water	convert to nitrate form with ion exchanger	Iodide	D	[264]
Molybdenum	convert to molybdenate, dilute 1 + 1 with methanol, titrate with $Pb(ClO_4)_2$	Lead	T/Pb^{2+}	[272]
Nickel	establish pH of Ni^{2+} solution at 8.5–10.5 with NH_4OH, add CuEDTA	Copper	T/EDTA	[270]
Nitrate				
air	rinse collector with 10^{-2} M AgF	Nitrate	D	[435]
foodstuffs	mix with water	Nitrate	D,S	[410, 411]
plants	extract with Ag_2SO_4 solution, filter	Nitrate	D,S	[412, 413, 414]
sea water	reduce to NH_3	NH_3	D,S	[264]
soil	stir sample in water, filter through acid-free filter, if not measured within half an hour add phenylmercuric acetate as preservative	Nitrate	D,S	[405–409]
water	no preparation with low chloride content	Nitrate		[415]
waste water	precipitate Cl^- with Ag_2SO_4 or reduce to NH_3	Nitrate/NH_3	D,S	[264, 404]
Nitrilotriacetic acid	treat with NH_3-NH_4OH buffer	Copper	$T/Cu(NO_3)_2$	[416]
Nitrite	mix sample 1 + 1 with a bromine-containing acetate buffer and measure resulting bromide	Bromide	S	[264]
Oxalate	pH 4.5–9.5, dilute 3 + 2 with p-dioxane	Lead	$T/Pb(ClO_4)_2$	[417]
Oxidizing atmospheres	absorb in pH 7 orthophosphate buffer + 5 x 10^{-5} M Iodide, measure decrease in iodide	Iodide	S	[418]
Perchlorate				
explosive	burn in Parr bomb, absorb gases in hydrazine sulfate solution, add HNO_3 and ethanol	Chloride	T/Hg^{2+}	[419]
solid fuel	dissolve sample in water	ClO_4^-	T/tetraphenyl arsonium chloride	[420]
Phosphate	add excess $La(NO_3)_3$ and back titrate with NaF	Fluoride	B/NaF	[264]
Potassium				
ATP solutions		Potassium	D	[396]
fertilizer		Potassium	S	[264]

serum	no special preparation	Potassium	D	[397, 248, 430]
soil	extract with sodium acetate solution	Potassium	D	[264]
wine	dilute 1 + 9 with water	Potassium	D	[264]
Rubidium	pH value 4 units above pRb value	Cation	D	[432]
Selenium	treat Se^{2-} solution with antioxidation buffer	Lead	T/$Pb(ClO_4)_2$	[264]
Silver				
fixing		Ag_2S	S	[421]
galvanic	dilute 1 + 99	Ag_2S	S	[264]
traces	dry $NaNO_3$	Ag_2S	D	[272]
Sodium				
boilers	establish pH 10–11 with NH_3 gas	Sodium	D	ASTM2791-69T
	industrial flow-thru measurements with diisopropylamine	Sodium	D	[398]
glass	decompose with lithium tetraborate, neutralize with HCl, pH 8–9 with NH_4OH	Sodium	D,S	[349]
paper industry	dilute, take aliquots with $(NH_4)_2CO_3$	Sodium	D,S	[401]
pharmacy	dissolve sodium salicylate tablets	Sodium	D,S	[403]
sea water	calibrate with synthetic sample	Sodium	D,S	[264]
serum	no preparation	Sodium	D	[82, 248]
skin	wet surface; use flat membrane electrode	Sodium	D	[400]
soil	extract with water or $CaCl_2$	Sodium	D	[264]
urine	dilute 1 + 19 with Tris buffer (pH 8)	Sodium	D,S	[264, 402]
wine	dilute 1 + 9 with dil. NH_4OH	Sodium	D,S	[264]
Sulfate				
soil	dilute 1 + 1 with methanol	Lead	T/Pb^{2+}	[422, 423]
galvanic	add sample to water + p-dioxane + Pb^{2+} solution, measure decrease in Pb^{2+} concentration	Lead	S	[272]
Sulfide				
cigarette smoke	absorb in 2 M NaOH, 0.2 M ascorbic acid, dilute 1 + 1 with water	Ag_2S	D	[428]
paper industry	add excess Cd^{2+} and measure decrease in Cd conc.	Cadmium	S	[426]
proteins	dissolve in water or dil. HNO_3, reduce to thiol with $NaBH_4$, add EDTA, warm, acidify to pH 3	Ag_2S	T/$AgNO_3$	[427]
water	dilute 1 + 1 with conditioning solution (antioxidizing)	Ag_2S	D	[264]

Table 8 (continued)

Sulfite	acidify to pH < 0.7 with $HClO_4$	SO_2	D	[168]
Sulfur				
organically bound	burn sample in oxygen bomb, absorb gases in $NaNO_3$ solution, pH 4–6.5 with NaOH, add p-dioxane	Lead	T/Pb^{2+}	[424, 425]
Sulfur dioxide				
air	absorb in water, add 1 + 10 SO_2 buffer	SO_2	D	[264]
water	as above	SO_2	D	[264]
wine	see section 3.5.5	SO_2	D	[168]
Tellurium	treat Te^{2-} solution with antioxidation buffer	Lead	$T/Pb(ClO_4)_2$	[264]
Thallium	Tl^+ solution, pH > 7	Cation	D	[432]
Thiocyanate		SCN^-/Cu	$D,T/Cu^{2+}$	[264, 224]
Thiol		Ag_2S	$T/AgNO_3$	[264]
Thiosulfate		Ag_2S	$T/AgNO_3$	[224]
Thorium	titrate Th^{4+} solution with NaF	Fluoride	T/NaF	[224]
Tungsten	dilute WO_4^{2-} solution 1 + 1 with p-dioxane	Lead	$T/Pb(ClO_4)_2$	[272]
Uranium	titrate UO_2^{2+} solution with NaF	Fluoride	T/NaF	[224]
Vanadium	treat V^{2+} solution with Cu EDTA	Copper	T/EDTA	[264]
Water Hardness	establish pH	Divalent	D	[429, 430]
Zink	set pH at 10 with NaOH, add CuTEPA	Copper	T/TEPA	[404]
Zirconium	titrate with NaF	Fluoride	T/NaF	[224]

Key: D = direct potentiometric, S = standard addition or subtraction, T = titration, B = back titration

In these bio-related fields, it is primarily the pH, pNa, pK, pCa, pCl or pF electrodes which are employed. Depending upon whether extra- or intracellular ion activities are to be determined, different electrode constructions are utilized. Measurements in extracellular fluids (whole blood, serum, etc.) are usually carried out with the conventional macroelectrodes. Anaerobic conditions are harder to maintain with this method. *It is expedient to thermostat the entire sample beaker and electrode pair set-up. For more accurate pH and pCa determinations the air space above the sample must also be equilibrated with an O_2-CO_2 mixture,* to establish the same partial pressure of CO_2 as in the original sample environment. This expense can be spared by working with microprobes under anaerobic conditions with small surface areas (low CO_2 loss). In such cases the commercially available flow-thru cells or capillary cells are used. For ion activity measurements *within the cell* special microelectrodes with tip diameters about one micron are employed, so that the cell membrane is not damaged too much. To date, such electrodes are available only for pH and pNa measurements (Transidyne, U.S.A.).

6.1.1 Measurements in Extracellular Fluids

Here again two cases can be distinguished. With *in vitro* measurements there is no great difference in practical techniques as compared to other samples (see Chap. 5), aside from the complexity of the sample matrix. In contrast to other fields of application, however, in biological systems the danger always exists that the sampling process may influence the physiological equilibrium. The consequence of such a disturbance is that no matter how accurately an analysis is carried out, the actual variable of interest is not measured. This problem is reduced with *in situ* or *in vivo* measurements; however, here problems of sterilization and electrode calibration come up.

6.1.1.1 *In vitro* Measurements

The following points should be noted: In extracellular fluids Na, K and Cl exist predominately in free, uncomplexed form. The analyses obtained with ion-selective electrodes are in good agreement with those carried out with other techniques (flame photometry, quantitative analysis, colorimetry). Calcium, however, shows an entirely different behavior in this regard. For example, the total calcium content of human serum as determined by atomic absorption is about 2.50 meqCa/l. If a potentiometric measurement (direct potentiometry) is subsequently made with a calcium-selective electrode, a calcium content of only 1.10 meqCa/l is found. This fraction of only 46% of the total calcium content represents the portion of calcium present in the form of free ions, and agrees within experimental error limits with the results of the frog heart method. The portion of bound calcium can be further subdivided. If atomic absorption is used to analyze a sample of dialyzed, protein-free serum sample, a total calcium content of 1.50 meqCa/l results, whereas the level of ionized calcium determined with the calcium electrode remains the same. This means that about 1.00 meqCa/l is strongly bound to proteins (CaProt). The dialyzable, but nonetheless bound portion of calcium of about 0.4 meqCa/l must be bound to some complexing agent (CaR). The levels of hydrogen carbonate, sulfate, citrate, lactate and phosphate present in serum are completely sufficient to account for this portion. Clinical experience has shown that the level of ionized calcium is notably constant in a healthy human. Moore [244] found higher calcium activities only in patients with specific cancers. Further, he was able to show that about 81% of the CaProt was bound to albumin, and only about 19% to globuline. *Thus, if the variation in total calcium content is determined with the help of atomic absorption, one more or less detects variations in the albumin level.* If only the total calcium content is determined, without regard for the albumin content, then a diagnosis might easily overlook a case of hyperparathymoism. In this condition, the total calcium content falls in the normal range, even with a low albumin level, but the level of ionized calcium is always significantly higher [244]. Moore found in addition, that the CaProt ratio in cancer patients was 66.5% Ca-Alb. and 33.5% Ca-Glob., a significant deviation from the normal value. The calcium binding ability of proteins decreases with lower pH values, apparently because hydrogen ions compete with Ca^{2+} ions for protein binding sites. The result is higher free calcium ion levels at lower pH values. *This is important in practice, since in order to obtain reproducible results, one must always work at the same pH value.*

6.1.1.1.1 Sample Preparation

Most errors are made in sample preparation. In potassium analyses of serum samples, a dependence of the total potassium content on the speed of blood sampling is found. Thus it is not always the analysis which is responsible for discrepancies between different control runs. When collecting blood samples turbulent currents and partial vacuums (too great a suction by the syringe) should be avoided. The resulting change in osmotic ratios can easily lead to a hemolysis with a corresponding increase in potassium level (the potassium level in intracellular fluids is about 30 times higher). Veins lying near the surface should be chosen for puncture, and, in case of obstructions, pumping motion with the hand should be avoided. Otherwise a higher potassium level than that in the usual extracellular space results in serum near the puncture [245]. Obviously, the freshly obtained blood sample should be worked up immediately (separation of the serum from the erythrocytes), or a whole blood analysis with ion-selective electrodes should be carried out at once. It is also clear, that hemolyzed serum cannot be used for the measurement of extracellular ion activities. Even for pure serum analyses fresh samples should be used. If storage is unavoidable, the samples should be cooled to freezing. *For whole blood analysis, one should test to find out whether the usual heparin addition partially complexes the ion to be measured. An approximately 5% diminution in free calcium ion activity has been observed upon addition of 50 units of heparin to a 1 ml sample*[1]. *This is to be taken into account when making comparisons between whole blood and serum analyses. A similar influence has been found on ion activity measurements due to the pH of the sample.* To avoid pH changes in the sample due to loss of CO_2, they should be handled under anaerobic conditions and stored under oil. For "pool" analyses with normal macroelectrodes, the desired pH should be established with an O_2-CO_2 mixture. Since this pH value is temperature dependent, the temperature at which the measurement is carried out also plays a large role, especially in pCa determinations.

[1] With 4.4 I.U./ml only negligible binding was found [458].

The developments of Beckman, Orion and Technicon (Analyzers for Na^+, K^+, Ca^{2+} and Cl^- ions) represent impressive examples of automation in clinical chemistry using ion-selective electrodes. Normally the result is reported in meq/l within about 3 minutes, which is accurate to about 2%. These automated instruments operate at optimum temperatures (37°C for Ca^{2+}, room temperature in other cases) and are self-standardizing with the help of an inner calibration solution. As an indication of the reproducibility of these direct potentiometric methods as compared to other techniques (for example flame photometry in the case of cations or coulometry for chloride), the correlation coefficients lie above 0.98 in all cases [246, 247]. The correlation coefficients between whole blood and serum samples are 0.961 for Na^+, 0.962 for K^+, 0.976 for Ca^{2+} and 0.991 for Cl^- [248].

6.1.1.1.2 The Indicating Electrode

With measurements in biological media, a protein layer often builds up on the electrode surface, influencing the potential (by about 1 to 4 mV). This effect is most notable with the first measurement in protein-containing solution following calibration in a pure water electrolyte solution. For this reason the indicating electrode should be brought into contact with the medium to be used before the usual calibration curve measurements. Some users conclude the protein solution measurements with a number of measurements of a standard solution, which in addition contains the main matrix ions of the sample (about 140 meq/l NaCl, 5 meq/l KCl and 1 meq/l $MgCl_2$) and then average the standard solution potentials. Others add trypsin to the aqueous standard solution, so that the protein layer automatically dissolves during the usual standard measurements. In this case, however, one should check (see Sect. 5.7) to make sure that the particular ion measured is not complexed by trypsin, thus lowering the standard solution activity and resulting in inflated values for the sample results. Occasionally, a previously, carefully analyzed serum is used as a reference standard. This represents a well buffered ion system and should be used with very small measured ion contents. In addition, this procedure leads to minimal change in the liquid junction potential.

6.1.1.1.3 The Reference Electrode

One problem with biological measurements is keeping the liquid junction potential between the reference electrode electrolyte and the sample solution constant. Most disagreements between blood pH measurements can be traced to fluctuations in the liquid junction potential. To avoid clogging of the diaphragm by albumin precipitation at the 3 M KCl reference electrode solution/albumin containing sample solution contact points, free bordering electrolyte solution contacts (see Sect. 2.3.4) are used almost exclusively. With many capillary set-ups the sample solution contacts the 3 M KCl reference solution through a thin (about 1 mm I.D.) plastic tube. Khuri et al. [226] used an equilibrium dialysis solution as salt bridge electrolyte. A sodium-selective electrode was successfully used as a reference electrode for blood pH measurements. Above all, of course, the sodium content of the test solutions must not change in this case.

6.1.1.2 *In vivo* Measurements

With the help of ion-selective electrodes and the proper calibration, the ion activity at the indicating electrode position can be monitored directly and continuously. According to Friedman [249], who was the first to carry out continuous blood pNa and pK measurements, one can distinguish between static and dynamic flow measurements. Static measurements are those such as ion activity measurements in the stomach or on the surface of the skin. *In vivo* flow measurements are primarily ion activity measurements in bloodstreams, either in the flow circuit of a heart-lung machine, or also post-operatively by means of vein shunts.

To date *in vivo* blood flow measurements of various ions have only been carried out sporadically. This is due to the problems associated with this type of measurement. Since ion-selective electrodes can usually tolerate temperatures only as high as about 50°C or just slightly above 100°C, the necessary sterilization is not so simple. *With the newer liquid ion-exchanger and ion-carrier electrodes, too little is known about the toxicity properties of the active phases.* The blood-compatible glass membrane electrodes often collect albumin deposits on the electrode surface during flow measurements, which increase with time and cause the electrode potential to drift. *Sometimes blood clots will also form, which are perilous to the research subject.*

Nonetheless, there is great interest in the continuous monitoring of blood pH, pNa, pK and pCa. Perkins et al. [250] report on the importance of maintaining a specific calcium ion level in the blood circuit of a heart-lung machine. Here the danger exists that the free calcium ion level may be lowered too much (complex formation) by the usual citrate addition. If this is counteracted by the addition of more calcium, too high a calcium level may result. Either occurrence can cause the heart to stop.

6.1.2 Measurements of Intracellular Ion Activities

As specific sensors, ion-selective electrodes can contribute a great deal to the understanding of the mechanism of the rapid activity changes of a few ions across the cell walls connected with the conduction of nerve impulses (information transmission). Microelectrodes have been manufactured which are suitable for the *in situ* measurement of K^+, Na^+, Ca^{2+}, Mg^{2+} and PO_4^{3-} activities. The earliest investigations [251, 252] showed that the potassium level inside the cell, at about 150 meq/l, was very high compared to that outside the cell, around 5 meq/l, and that during cell activation some ions apparently migrate against their activity gradients (potassium-sodium pump). One can imagine the important role played by ion-selective electrodes in the search for the energy-liberating process necessary to complete this cycle.

In order to minimize damaging or interfering with the cell to be investigated *puncture electrodes* with tip diameters of < 1 micron must be used. This corresponds to the diameter of the microcapillary pipettes commonly used in physiology, whose insertion into cell interiors of dimensions as small as 0.5 μl has been shown to exhibit no significant interference with the cell function. Extreme care must be taken to insure that the active part of the ion-selective electrode only comes in contact with the inner fluid to be measured. Otherwise a mixed potential is measured, which depends on the activities of ions in both intra- and extracellular fluids. How such glass membrane electrodes with 5–10 μ long active zones are constructed from H^+-, Na^+- and K^+-sensitive glasses has been described in detail elsewhere [253]. A few possible construction designs are illustrated in Fig. 54. Hinke [252] has long used ion-selective glass membrane microelectrodes for intracellular sodium and potassium activity measurements. He found that almost all of the in-

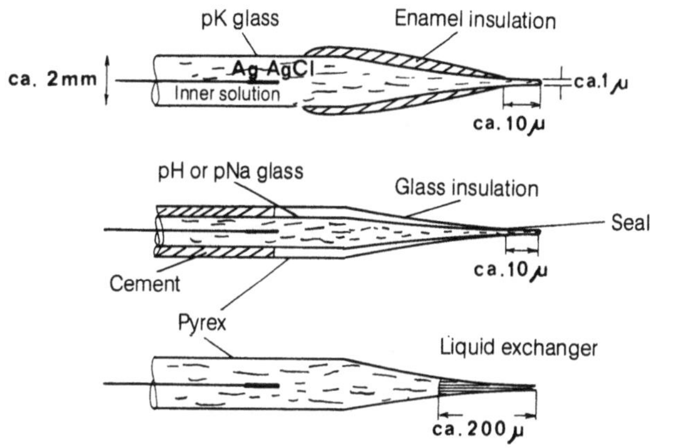

Fig. 54. Microelectrode constructions for intracellular measurements

tracellular potassium content existed as the free ion, as opposed to sodium, which is partially bound. Also to be noted is the fact that the activity coefficients of potassium and sodium differ to a greater extent than theoretically expected. Kostyuk et al. [67] found, for example, that in snail nerve cells the average potassium activity coefficient was 0.73, while that of sodium was 0.46. This is in agreement with the results of Hinke [252] and Lev [254]. There may be a number of reasons for such a large difference between two very similar ions. Some bare or hydrated sodium ions may be strongly bound to proteins. The presence of polyelectrolytes can also cause the individual activity coefficients to differ greatly from the mean, since the main assumption of the Debye-Hückel approximation (that of a radially symmetric point charge, see Appendix) is no longer valid. It was demonstrated that the mean sodium ion activity coefficient changes drastically with the geometric form of the polyelectrolyte (monomer → polymer) [67].

Since the introduction of ion-selective electrodes with liquid ion-exchanger phases there are more possible microelectrode designs, as the fluid active phase can easily be placed in the tip of the capillary pipette electrodes usually employed in investigating biological cells.

6.1.2.1 Preparation of Ion-Selective Microelectrodes

A commercially available hard-glass capillary (borosilicate glass, Duran, Pyrex) of about 1.0 to 1.5 mm outer and 0.4 to 0.5 mm inner diameter is used to start with. After thorough cleaning (hot, dust-free ethanol vapor) it is drawn to a micropipette with an inner tip diameter of less than 1 μ with the aid of a glass drawing device, commonly used in physiology. The tip zone which will eventually hold the ion-selective liquid ion-exchanger phase is siliconized for a stretch of 200 to 300 μ by dipping in a 5% solution of dichlorodimethyl silane in carbon tetrachloride. The excess hydrophobic coating agent is blown out with a stream of dust-free air, and the micropipette is stored in a dust-free location with the tip upward for 24 hours to dry. Bubble-free filling of the pipette with the inner solution corresponding to the particular ion-exchanger phase to be used (see Chap. 4.1) is accomplished by first filling with methanol, which is then carefully displaced with the special inner solution from the larger end. An Ag/AgCl element is chosen for the reference electrode, so a chloride salt of the corresponding measured ion is used for the inner solution electrolyte whenever possible. If this is not the case, then a certain amount of KCl or NaCl (corresponding to 0.1 M solution) must be added to the inner solution to establish a defined potential at the reference element. The inner solution measured ion activity should roughly correspond to that in the samples to be measured, because very large differences in its activity across the ion-exchanger phase can lead to deviations from Nernstian behavior [68]. A thin, chloridized silver wire shielded at least as far as the liquid surface is immersed in the inner solution as a reference electrode (for preparation see Sect. 2.4.1.1). Shortly before the measurement the tip of the appropriately filled micropipette is dipped in the ion-selective exchanger solution. Normally the organic exchanger solution is immediately taken up into the hydrophobic zone of the tip (about 0.2 mm). If not, this process can be aided by applying a partial vacuum to the other end of the micropipette. To avoid any evaporation of the inner solution, its surface is covered with a layer of oil. Experience has shown that of the commercial ion-selective exchanger phases available today, those from Corning show a higher conductivity than those from Orion. For preparing these special microelectrodes, exchangers which exhibit a lower electrical resistance are preferred. Directions are also given in the literature for the preparation of two- to four-fold microelectrodes [255,459]. In these procedures one starts with special glass tubes which are divided into two to four compartments. With proper drawing of the capillary these compartments remain intact. Thus with a tip diameter of only 1 μ, a reference electrode can be included in a second compartment, or in the case of four-fold microelectrodes, three ion-selective electrodes and one reference electrode! Filling these minute capillaries is very difficult. To insure that only one capillary compartment is filled with a given ion-selective phase, a partial vacuum is applied to one chamber, while those remaining are all supplied with positive pressure [256]. According to another procedure, the ion-selective phases are introduced by injecting into the larger openings and are subsequently centrifuged into the tip [257].

With such tiny cross-sections, the resistances of these electrodes can lie in the vicinity of 10^{12} Ω, so that technical problems again come to the foreground (insulation, time constant, electrostatic induction, etc.). These will be briefly discussed in the following sections.

6.1.2.2 Guarded Signal Wires for Extremely High Ohmic Electrodes

If the electrical resistance of the ion-selective electrode becomes so large (also with thick-walled glass membrane electrodes at lower temperatures) that it begins to be of comparable magnitude to the insulation resistance of the electrode cable material, then only a special guarded signal wire can be of any help. In this case the shielding (see Fig. 40) is not at the ground voltage as usual, but rather at that of the signal-carrying cable core. This can be easily realized by applying a corresponding voltage across the shielding and the instrument leads as shown in Fig. 55. In this way the voltage drop between cable core and shielding is virtually zero and, necessarily, also the leakage current between the two, so the electrode potential is measured without error. In order for the shielding to maintain its full effectiveness (diverting inductions to the ground) a low ohmic voltage source ($<$ 1 Ohm) must be used to adjust the voltage of the shielding to that of the cable core. Some electrometer amplifiers also contain another special output for this purpose; the so-called "unity gain" output. At this instrument plug a voltage exactly corresponding to that of the respective measured potential is provided (very low ohmic).

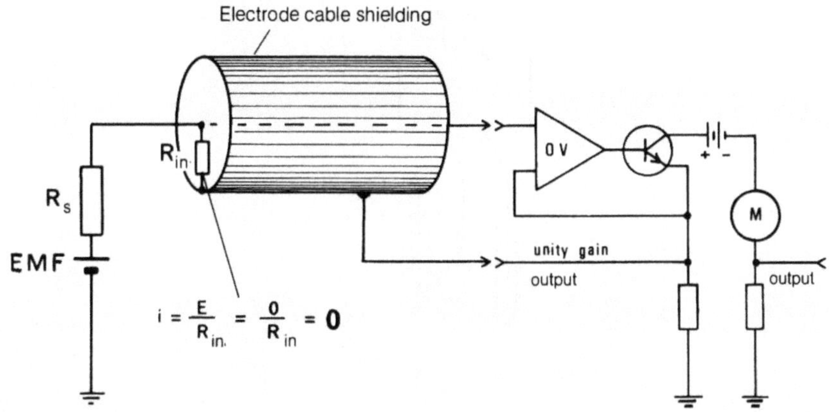

Fig. 55. Principle of a guarded signal wire in case of a source resistance of the same magnitude as the cable insulation resistance

6.1.2.3 Reference Microelectrodes

The Ag/AgCl micropipette reference electrodes usually employed in physiological measurements of cell membrane potentials can be used as reference electrodes for intracellular ion activity measurements as well. The salt bridge electrolyte must be chosen with care so that the intracellular measured ion concentration is not influenced.

There are two possibilities concerning the positioning of the merely 1 μ diameter reference electrode tip in the cell structure:

a) It can be introduced into the cell interior exactly as the indicating electrode. To do this (if a multiple construction is not available), the reference electrode tip is fixed next to the indicating electrode with plasticine, sealing wax, etc. under a microscope, so that the distance between the two amounts to only about 1 μ. They are then inserted into the cell with the help of a micromanipulator. According to previous calibration, which should also be repeated between measurements, the measured ion activity can be obtained directly from the EMF between the two electrodes. If an additional reference microelectrode is positioned in the extracellular region near the cell, then the rest potential of the cell can be simultaneously registered between the two reference electrodes (see Fig. 56). This arrangement is of great advantage because it is then easy to verify whether the reference electrode is in fact within the cell interior. Disturbances in cell function are also recognized at once.

b) If the minute size of the cell to be punctured prohibits the introduction of two microelectrodes into the cell interior, the intracellular ion activity measurement can be carried out with a reference electrode positioned outside the cell. In this case the normal rest potential (average value) must be taken into account during the activity calculation. Naturally, the appropriate "tip potentials" are also to be considered (details in [253]).

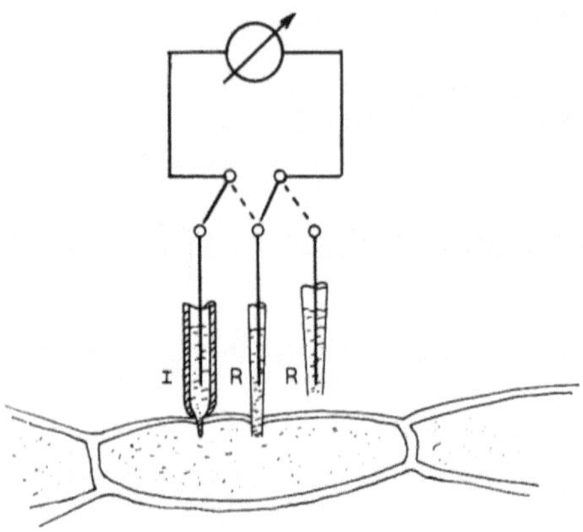

Fig. 56. Positioning of the microelectrodes in the cellular structure

6.1.2.4 Special Amplifiers for Microelectrode Applications

One can also try to minimize the insulation problems which arise when working with ion-selective microelectrodes by completely doing away with the cable insulation. Especially with microelectrode pairs, the silver wires of the two reference elements can be soldered directly onto the high ohmic input contacts of a miniaturized operational amplifier. For shielding considerations a small sheet metal case, used to house the operational amplifier, is guided down just over the electrolyte filled electrode shafts. A teflon plate can serve as a spacer and mounting plate. On the other side of the amplifier case a micromultiplug is positioned for attaching the supply voltage and for further transmission of the now low ohmic measured signal. Walker [258] used a Varactor-Bridge operational amplifier (Analog Devices, Model 311 C) with an input impedance of about 10^{14} Ohm as an impedance transformer. For measuring the low ohmic voltage after the operational amplifier he used a common digital voltmeter. With this simple impedance transformer he reproducibly attained a stability (including electrode drift) of better than 1 mV over the course of many hours. Since the electrode resistances cannot be substantially reduced, a small input capacitance must be provided for in the interest of a rapid indication. In light of the fact that a cable capacitance of 100 pF per meter of shielded cable must be reckoned with, both electrode wires should be lead to the high ohmic amplifier inputs through the shortest, unshielded (the amplifier housing serves this purpose) route, and should be insulated from each other only by the teflon plate. To reduce the input capacitance of the electrometer amplifier input even further, the "C-neutralization" procedure (Fig. 57) can be utilized. Neher and Lux [259] attained stable final values after only about 20 msec with this method. It is advantageous to use a Faraday cage for all measurements with these microelectrodes.

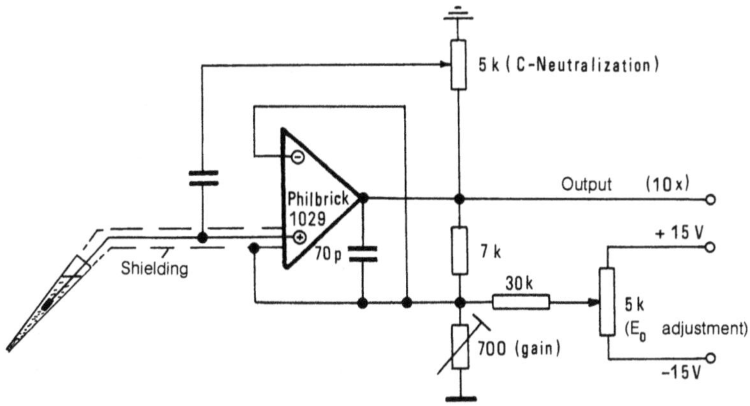

Fig. 57. Operational amplifier circuit with C-neutralization from [259]. A type 1029 unit of Teledyne Philbrick constructed with the MOSFET technique is used; this exhibits an input impedance of 10^{13} Ω for an input capacitance of 3 pF

6.2 Continuous Measurements in Industry and Environmental Research

One of the great advantages of these new electrochemical sensors is their ability to continuously provide *an electrical signal proportional to the measured ion activity* or concentration without complicated signal transforming devices (mechanical-electric or optical-electrical transformers). The logarithmic relationship between the measured ion activity and the EMF allows operation over many orders of magnitude in concentration with the same relative accuracy. This dynamic working range together with their simple, sturdy construction and maintenance-free operation (liquid exchanger electrodes with monthly regeneration excepted) make these ion-selective electrodes uniquely suited for continuous measurements in process control and as monitors for environmental protection. In principle, the same considerations apply here as with the discontinuous analytical techniques (see Chap. 5) concerning activity or concentration calibration, direct or indirect procedures and potentiometric or titrimetric techniques. In the final calculation formulas, the volume is merely replaced by the corresponding flow rate.

To date, automatically calibrating monitors have been offered on the market for the continuous indication of:

Aluminum	Fluoride	Sodium
Ammonia	Hydrogen Sulfide	Sulfate
Ammonium	Hypochlorite	Sulfide
Bromide	Iodide	Sulfite
Cadmium	Lead	Sulfur Dioxide
Chloride	Nitrate	Total Acid
Chlorine	Nitrite	Total Base
Copper	Oxidizing Agents	Water Hardness
Cyanate	Reducing Agents	
Cyanide	Silver	

These all exhibit detection limits below 1 ppm, and are offered by the firms:

Bran & Lübbe, Hamburg (Ionometer), Electrofact, Netherlands (Na monitors for boiler feed water supervision), E.I.L., Richmond/Surrey, England (Water quality monitors, Automatic selective ion monitor), Foxboro, U.S.A., Orion, Cambridge, Mass., U.S.A. (Series 1,000 monitors), Polymetron, Hombrechtikon, Switzerland (Condimat system), Union Carbide, U.S.A. (Water monitors) and others.

In the fields of process control, waste water monitoring, environmental protection and pollution control, the concentration is of interest rather than the activity indicated by the electrode. This means that the sample stream must be conditioned ahead of time. Through the addition of a constant ratio of the conditioning solution appropriate for the particular application, the concentration itself can be directly monitored. The *conditioning solution* has the following tasks, as discussed in Chap. 5.1.2:

1. Keeping the ionic strength constant by means of an ionic strength adjustor (1–5 M solution of an inert electrolyte), so that the resulting constancy of the activity coefficients can allow direct concentration indication;

2. Providing the optimum pH value with a suitable buffer system – some electrodes indicate H^+ also, and some measured ions react with H^+ or OH^- to form undissociated, volatile or insoluble precipitate species;

3. Setting free any bound measured ion through an exchange reaction with a complexing agent appropriate for the particular case;

4. Binding any interfering ions with special reagents (such as removing oxygen with ascorbic acid in the measurement of sulfide traces);

5. Providing a constant amount of the ion actually detected in the case of an indirect titration.

Unfortunately there are no generally valid rules for industrial measurement techniques. The commercially available instruments are usually flexible enough, however, to be compatible with the flow schemes in other special applications. The manufacturers also have the most recent, detailed procedural directions available for many of the substances listed in Table 7. As a representative example, the problems associated with automating one particular determination will be discussed here. First, however, the flow-thru cell itself and some error-determining parameters will be briefly discussed.

6.2.1 Flow-thru Cells

The flow-thru cell constructions already long employed for industrial pH measurements can also be used with ion-selective electrodes. Here the reference electrode is positioned further along in the flow direction than the indicating electrode to avoid interferences from outflowing reference electrode electrolyte. A single reference electrode can be used with a number of different ion-selective electrodes placed in the sample channel. If the individual electrodes are placed in the center of the sample stream, quick responses to activity changes can indeed be obtained, but streaming potentials (see Chap. 1.2) of up to 1.3 V, depending on the conductivity, sample channel diameter and flow rate, must be dealt with. Van den Winkel, Mertens and

Massart [436] reported interfering oscillations in the indication of up to 1,350 mV due to the streaming potential of a 5 x 10^{-5} M NaCl solution in a Technicon Autoanalyzer. These oscillations originated from the peristaltic pump used. As discussed in Chapter 1.2, the streaming potential can be reduced by increasing the conductivity of the sample solution. Where this is not possible due to poor electrode selectivity (such as the nitrate electrode), pumps should be used which pump evenly, or which let in air bubbles (removed shortly before the indicating electrode) in order to buffer the flow. According to the above authors, these oscillations could be reduced by a factor of 1:50 to 1:100 by using a grounded metal connection between the indicating and reference cell compartments. Another method to avoid streaming potentials is to use a different flow geometry. A small portion of the sample flow can be diverted to the electrodes as shown in Fig. 58. If streaming potentials do arise, one must at least be sure that a

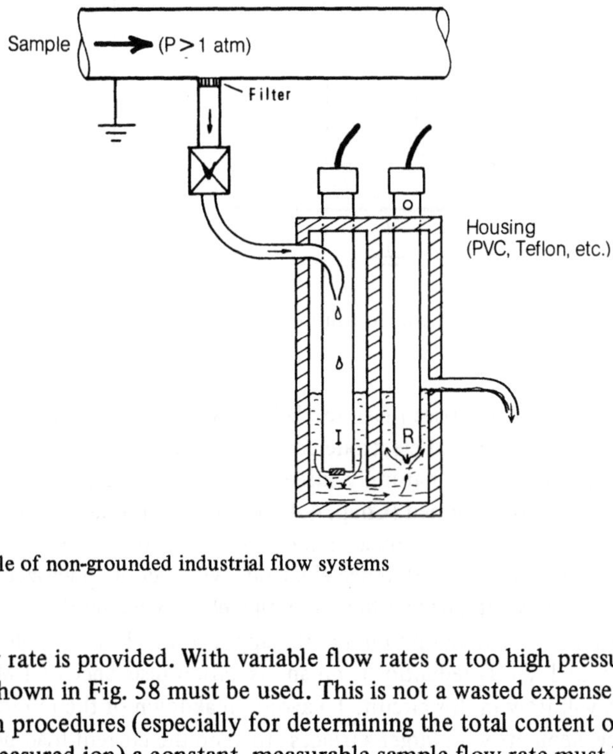

Fig. 58. Principle of non-grounded industrial flow systems

constant flow rate is provided. With variable flow rates or too high pressures a design such as that shown in Fig. 58 must be used. This is not a wasted expense, since for some addition procedures (especially for determining the total content of a partially complexed measured ion) a constant, measurable sample flow rate must be provided anyway.

Sample streams contaminated with particulate matter can be filtered before analysis, as shown in Fig. 58. Due to the self-cleaning effect of a horizontal flow, such filters can even be used with highly contaminated samples for over a month without changing the filter [437].

6.2.1.1 Influence of Grounding

In chemical technology, pipelines are often made of steel or cast iron and are usually in contact with the ground. Thus direct measurements of sample streams in such lines are made in grounded solutions. In this case an amplifier with sufficient insulation between the shielding ground (= reference electrode potential) and the mains ground must be chosen. As the schematic diagram in Fig. 59 attempts to illustrate, with insufficient insulation the danger arises that the EMF between the reference electrode and the pipe wall behaving as a metal electrode can cause a definite current to flow in this circuit. If this current, which is naturally susceptible to other stray currents due to the size of the ground loop formed (Chap. 4.4.4), becomes too large (a few μA), then the conditions for a practical EMF measurement (constant reference electrode potential) are no longer fulfilled. The reference electrode will become polarized, even though almost no current is flowing through the indicating electrode circuit. Currents in the μA range are not at all uncommon in industrial applications. For example, with a possible EMF of about 1 V (O_2 reduction) and an insulation resistance of 1 MOhm, a current of 1 μA will flow in this circuit. To avoid breakdown of the reference electrode, special reference elements with larger surfaces (to keep the current load per unit area small) are used for measurements in grounded systems [438]. There are two other possibilities for reducing this current, both of which are based on effectively breaking this circuit. A battery-operated amplifier with a correspondingly larger insulation resistance can be employed. With the current integrated circuit technology the power requirements for a pH amplifier are so low that batteries can last for months. This circuit can also be broken elsewhere. Fig. 58 shows how the electrolytic contact to the main pipeline can be broken simply by using a flow cell constructed of a synthetic material (PVC, teflon, etc.).

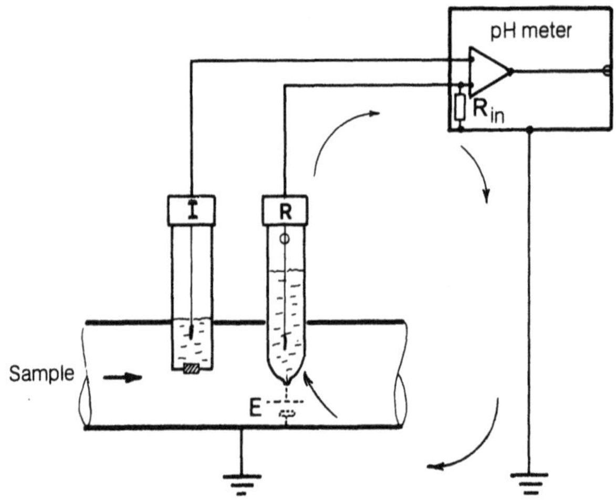

Fig. 59. Ground loop with grounded sample solutions

6.2.1.2 Influence of Temperature

All on-line monitors are accompanied by the problem of how to eliminate the temperature influences upon the EMF measurement. Since contradictory viewpoints concerning this problem exist in the literature and some manufacturers recommend completely inappropriate temperature compensations, the subject will be treated in some detail here.

In order to understand the influence of temperature on the EMF measurement we must take a closer look at Fig. 32. With a change in temperature each individual Galvani potential in this circuit changes. Some of these Galvani potentials follow temperature changes in the sense of the Nernst equation. The temperature effects on $\Delta\phi_1$, $\Delta\phi_6$, $\Delta\phi_2$ and $\Delta\phi_5$, which are located in opposing positions within the circuit, can only compensate for one another (ignoring the temperature dependence of the liquid junction potential) in a highly symmetric cell design, i.e. when both reference electrodes are of the same type, so that $\Delta\phi_1 = \Delta\phi_6$ and $\Delta\phi_2 = \Delta\phi_5$. In this case, for identical inner and outer (sample) solutions, and again ignoring the liquid junction potential, an EMF of zero is expected. In this special case $\Delta\phi_3 = \Delta\phi_4$ at all temperatures, and the cell shows absolutely no temperature dependence. Only when the measured ion concentration in the outer solution deviates noticeably from that of the inner solution is an EMF detectable, whose temperature dependence can be approximately (only about 90–98% of the theoretical value) described by the temperature dependent term of the Nernst equation.

If the measured EMF values of such an ideal system are plotted as a function of both temperature and measured ion activity (EMF vs. log a_M), then a series of isotherms result which have different slopes, but which all intersect at an EMF value of zero. In such an ideal case automatic temperature compensation is not difficult. In order to obtain temperature independent activities, the sensitivity of the instrument (mV per power of ten activity) need only be automatically adjusted for the appropriate measuring temperature. As a rule, a temperature dependent resistance (Pt 100) serves as a temperature sensor, which automatically changes the pH meter amplification by the appropriate factor.

In practice, however, even for simple pH measurements some of the assumed conditions are not met. Thus with glass membrane electrodes a so-called asymmetry potential is known, which is the potential still registered when inner and outer solutions are identical. The asymmetry potential of glass membrane electrodes depends on the thickness of the membrane, the manufacturing conditions, the pre-treatment, etc. and is subject to variations, which means the pH meter must be frequently calibrated. In this case the isotherm intersection point no longer agrees with the cell zero point. Here the automatic temperature compensation via changing the amplification can no longer be applied, since with this method the pivot point of the isotherms must lie at zero mV. In order to carry out a "proper" temperature compensation, the pivot point of the variable amplification must be set at the isotherm intersection point. Figure 60 indicates the situation which is more commonly found in practice. In this case an auxiliary voltage source must apply a voltage of magnitude E_{iso} to compensate for the isotherm intersection point before the measured signal is properly treated by the temperature dependent amplifier. EMF instruments which apply a "proper" temperature compensation via the isotherm intersection point are described in the DIN 19265 [439]. These amplifiers are constructed so that the coordinates of the previously ex-

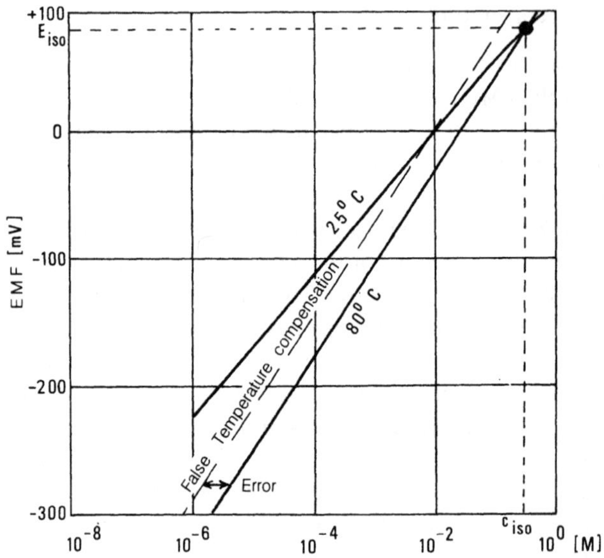

Fig. 60. Measurement errors due to false temperature compensation

perimentally determined isotherm intersection point can be entered on two control knobs, E_{iso} and c_{iso}. In contrast to pH measuring cells however, with ion-selective electrodes E_{iso} values of ± 800 mV and c_{iso} values between 1 and 10^{-10} M must be dealt with, so the two variable voltage sources must be dimensioned accordingly. Oehme [356] found, for example, with a cyanide-sensitive cell an E_{iso} value of ~ –420 mV and a c_{iso} ~ 3 x 10^{-2} M CN^-, and was able to show that the Condimat system of Polymetron (Switzerland) with an amplifier according to DIN 19265 completely compensated for temperature fluctuations in a practical cyanide control cell.

Due to the complex construction of an ion-sensitive electrochemical cell (Fig. 32) the position of the isotherm intersection point cannot be theoretically predicted. It must be empirically determined for each individual electrode cell. Caution must be exercised here to insure that the conditions of this determination match those later to be employed in measurements. (To reproduce the same temperature gradient along the electrodes, even the air temperature should correspond to that which will be used for the actual measurements.)

The isotherm intersection point is best determined in a flow cell of the same design as that at the monitoring location. The rate of heat transmission in the temperature sensor used (usually a Pt-100 resistance) should resemble that of the electrode, so that rapid, temporary temperature fluctuations of the sample solution do not result in overcompensation.

The contradictory statements concerning the validity of automatic temperature compensation via the isotherm intersection point in the literature are based on primarily two arguments: It may happen that with more than two isotherms no single intersection point is obtained, but rather a diffuse intersection region, which naturally subtracts from the accuracy of the temperature compensation. In addition, with pH measurements there is a unique pH scale for each temperature, because the autoionization constant of water is itself temperature dependent (at 100°C, for example, the neutral point is about pH 6). In the last case a "proper" temperature compensation can indeed reduce electrochemical measurement errors, but does not change the fact that in principle, two pH measurements obtained at different temperatures cannot be meaningfully compared [17].

The fact that sometimes no definite isotherm intersection point is found rests on the unsymmetrical construction of the electrochemical cell. Since a chloridized silver wire in a chloride solution is usually used as an internal shunt electrode in ion-selective electrodes, a symmetrical cell construction requires that the outer reference electrode also be Ag/AgCl surrounded by a KCl solution of similar chloride ion activity. If, as is usual, a saturated KCl solution is used, then in addition to the Nernstian temperature dependence at ϕ_5, another dependence arises due to the varying solubility of KCl at different temperatures, unnecessarily complicating this relationship. If an exact calibration of the isothermal intersection point of a particular instrumental set-up has not been carried out, under certain conditions the error due to differing measurement temperatures can be kept small by using an indicating electrode with a refillable inner solution and constructing a symmetric circuit. Then the measured ion concentration in the inner solution must be adjusted so that the isotherm intersection point comes to lie in the mV region of the sample solution. If the cell is then calibrated with a standard solution which lies in the middle of the expected sample concentrations, then the measurements are obtained very near the isotherm intersection point and temperature variations of 30°C correspond to only a $< 2\%$ measured error.

In this way Light [439] was able to construct an on-line fluoride monitor for drinking water supervision. With temperature fluctuations between 20 and 45°C, errors could be kept under 4 to 10%, depending on the distance of the measured point from the calibration point.

The half-automatic temperature compensation entered by hand, frequently carried out in laboratory practice, also needs adjustments via the isotherm intersection point, but this is usually overlooked. For a desired accuracy of about 1% with a calibration solution differing in concentration from the sample solution by a factor of 5, the temperature potentiometer must be able to be set within an accuracy of 1°C. For a factor of 100 difference between sample and calibration solutions this tolerance is lowered to about 0.2°C. This illustrates how important these slope potentiometers are for accurate measurements. Unfortunately, they can often only be set to an accuracy of about 2°C.

If auxiliary chemical reactions are involved in concentration determinations using ion-selective electrodes (end point titration, standard addition or subtraction, indirect procedures), extreme caution is required with variable sample temperatures. This is because in addition to the electrochemical effects discussed, purely chemical phenomena may become problematic (temperature dependence of equilibrium constants, solubility products, complex formation constants and activity coefficients). With the low sample flow rates commonly encountered (a few ml/minute), thermostating the solution and sample cell with the help of a quickly responding proportional controller (Orion, Series 1,000) presents no problem.

6.2.1.3 Analysis Techniques with Flow-thru Measurements

Many ion-selective electrodes must be used within specific pH ranges due to direct H^+ ion indication or chemistry involving the measured ion (H^+ complex formation with the measured ion) and conditioning of the sample solution is often required for direct concentration indication. For these reasons it is advantageous to work with diverted streams having constant, known flow rates (see Fig. 58) whenever possible. The auxiliary pump for controlling the addition of a suitable conditioning solution can be mechanically coupled to the first pump maintaining a constant sample flow rate (for example a peristaltic pump), so that a constant feed ratio is maintained. If only a rough pH adjustment is necessary (acidic or alkaline, as with the sodium selective glass membrane electrode), this can be achieved through the introduction of a suitable gas (HCl or NH_3). *On the other hand a whole range of substances important in air pollution, such as HF, HCl, SO_2, NO, etc. can be specifically detected with ion-selective electrodes.* Direct potentiometric procedures can also be used for indirect analyses (see Chap. 5). In this case an ion which can be directly indicated and which reacts quantitatively with the substance to be determined is added to the conditioning solution instead of a complexing agent. In this way the aluminum concentration can be determined via the decrease in fluoride concentration. One must be sure, of course, that in this case the ion to be determined is never present in larger amounts than the indicated ion.

6.2.1.4 Flow-thru Measurements without Reference Electrodes

With an electrode arrangement such as that shown in Fig. 61 no reference electrodes are needed, only two identical indicating electrodes (same slope and same selectivity

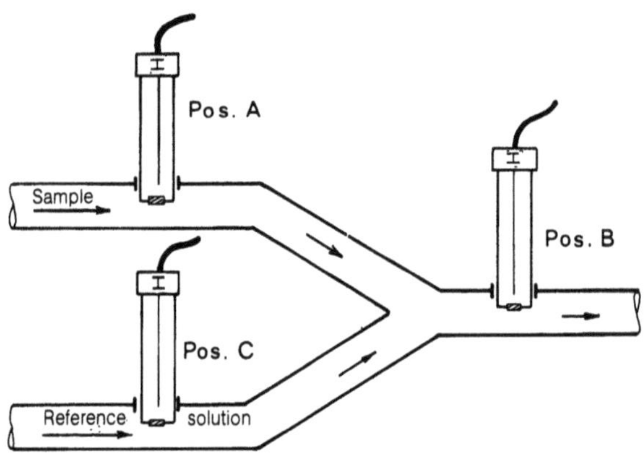

Fig. 61. Concentration cell construction for industrial flow-thru systems

behavior) at any two of the positions in the fluid streams illustrated. Such a construction has the advantage that reference electrode salt bridge diaphragms which can easily give rise to interfering effects can be done away with. The liquid junction potential at the junction of the two electrolyte streams can be neglected due to the rapid hydromechanical mixing, since differences in ionic mobilities do not exhibit as large an effect here as they do at static contact zones. Depending on the positions in the flow scheme at which the two ion-selective electrodes are placed, the methods described in Chap. 5 of direct potentiometry, standard addition and standard subtraction can be utilized here in a continuous fashion.

When replacing a reference electrode by a second ion-selective electrode, one must be especially aware of the fact, that with very large electrical resistances ($> 1\ k\Omega$) an amplifier with two high ohmic inputs (differential amplifier) must be used!

6.2.1.4.1 Direct Potentiometry (Positions A & C)

This design works on the principle of the concentration cell. The sample streams past position *A*. In order to be able to use the second indicating electrode in position *C* as a reference electrode, a constant single electrode potential must be provided for. This can be done by furnishing the comparison solution with a constant indicated ion activity. Care must be taken then that the flow rates are high enough that no backward diffusion of the indicated ion from the more concentrated to the more dilute solution can occur. In order to save chemicals, a dilute solution of the indicated ion is used for comparison in these cases. In order to avoid concentration instabilities when working in such highly diluted solutions (irreversible adsorption processes), it is advantageous to work with ion buffers. The activity or concentration is evaluated, as shown in Chap. 5.1, either with a calibration curve or by direct indication on an instrument which can itself be fitted to a previously determined calibration curve.

6.2.1.4.2 Concentration Determination through Standard Solution Addition (Positions A & B)

Here again the electrode potential at position A is determined by the measured ion activity in the sample stream. The electrode potential at point *B* in the flow diagram is determined by the sum of the measured ion activities in the sample and comparison solutions, provided that a constant indicated ion activity is supplied in the comparison stream and that the dilution upon combining the two streams is taken into account. In this case the EMF difference ΔE (see Chap. 5.4) can be read directly off of the amplifier. This corresponds to the change in activity of the sample solution. If the concentration of the comparison solution is known, this allows one to infer the *total initial indicated ion concentration.* Considering the flow rates, a volume per minute can be calculated, and these values can be used for calculations using the same formula (65) as that used in discontinuous processes. For direct registering of the output concentration corresponding to this ΔE and ΔC, an on-line operation using a computer must be employed. For small concentration ranges a simple linearized ΔE vs. C_0 calibration curve sometimes suffices. The same considerations regarding possible sources of error discussed in Chap. 5.4 apply here.

Naturally, in place of a comparison solution with constant measured ion activity, a reagent solution can be added which lowers the measured ion activity by a specific amount through precipitation or complexation of the indicated ion (subtraction method).

6.2.1.4.3 Indirect Concentration Determination (Positions B & C)

This procedure allows the determination of species which are not directly indicated by an electrode, but which react quantitatively with some indicated ion. The comparison solution contains an ion indicated by an ion-selective electrode in position *C*. The ion to be measured quantitatively combines (precipitate, complex) with this indicator ion. A lower activity of this indicator ion results at point *B*. Examples of this are the determination of Al^{3+} ions as already mentioned, of sulfate ions by means of two lead-selective electrodes and a $Pb(ClO_4)_2$ standard solution or phosphate ions with two fluoride-selective electrodes and a $La(NO_3)_3$ standard solution. The calculation corresponds to that of the subtraction method.

The accuracy of all these more or less direct potentiometric procedures can amount to within ±2 to ±5% for monovalent and ±4 to ±10% for divalent ions, depending on the interfering ion levels.

6.2.1.4.4 The Cyanide Monitor as an Example of Industrial On-Line Measurements

Due to the high toxicity of cyanide, a strict and "correct" supervision of potential emission sites (galvanizing, iron and steel industries) is of particular relevance in environmental protection. As Oehme [352] has already convincingly demonstrated, a multitude of difficult problems (which will not be gone into any further here) arise, especially for prolonged automatic recording, with photochemical as well as with the earlier electrochemical procedures (simple measurement of redox potentials or measurements made with amalgamated electrodes susceptible to aging). Compared to the batch control processes commonly used earlier, which are subject to relatively large subjective errors, considerable success has been attained in this special field of environmental protection through the development of suitable cyanide-sensitive electrodes. To date reliable cyanide monitors based on ion-selective electrodes have been offered by at least four different firms for routine industrial processes. The development of these monitors, including the suppression of undesirable interferences through sufficient "know-how" during the sample preparation and addition phases, is indicative of what can be expected for further adaptation of "laboratory methods" to industrial practice. This topic shall be briefly discussed to acquaint the reader with some of the problems associated with such development. The situation here is unfortunately not as simple as that of a laboratory pH measurement. A thorough understanding of fundamental analytical science is often necessary and sometimes years of experience and testing of optimum conditioning solution compositions are required. Thus it is understandable if the manufacturer is reluctant to disclose the composition of such a solution.

The Condimat System of Polymetron. This system is easily adapted for other determinations as well. The basis of this system is a cyanide-sensitive cell with an AgI/Ag_2S electrode with inner filling solution and a reference electrode with double liquid junction. In the interest of longer maintenance-free periods, both salt bridge electrolytes (inner: NaCl solution; outer: 2 M KNO_3) are fed from supply vessels positioned above them. Thus the hydrostatic pressure provides for a constant liquid junction potential and hinders penetration of the sample solution and clogging of the diaphragm at the tip of the reference electrode. As shown in Fig. 62, the sample to be analyzed and the conditioning solution (condisol) are drawn in by a two channel dosing pump

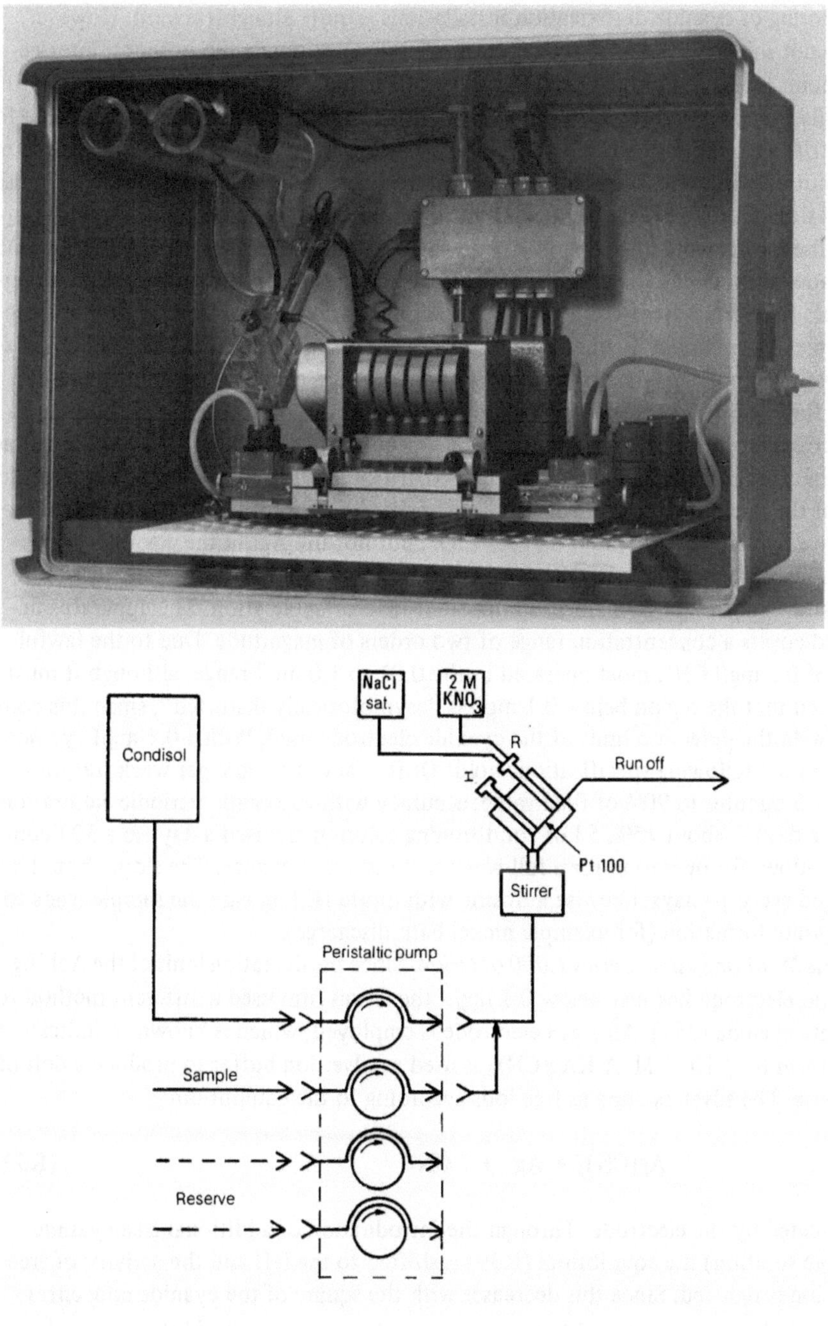

Fig. 62. Cyanide monitor (Polymetron). Above: Instrument with front plate removed; Below: Flow diagram

and added via a mixing chamber to the actual measuring cell in a flow-thru channel tipped at an angle of 45°. The latter inhibits air or gas bubbles from collecting near the electrode and causing unstable indications.

From the simple flow diagram it would appear that the automation of discharge monitoring of cyanide detoxication installations is quite straightforward. However, this is not so! First, the proper temperature compensation via the isotherm intersection point is realized in the Condimat system. Temperature fluctuations of about 30°C are fully compensated for with this system. In addition, a peristaltic pump suitable for industrial applications had to be developed. Last but not least, an optimum conditioning solution had to be found for this special application. This solution has to establish a pH value of > pH 11 (0.1 M NaOH) in the sample solution, since in more acid solutions the results would be too low due to the portion of CN^- present as HCN. In practice, however, making the sample so highly alkaline results in the precipitation of carbonates from the water hardness. These precipitates deposit on the electrode surface and quickly render any further measurements impossible. The addition of a complexing agent (polyphosphate) to the conditioning solution can be of help in this case.

A further interference arises if the cyanide detoxication process goes "too far on the active chlorine side". At hypochlorite concentrations > 10 mg/l, the AgI in the active phase is converted into AgCl, leading to drifting results and eventually to destruction of the membrane. For this reason the conditioning solution must also contain a selective reducing agent which reduces ClO^-, but not the AgI in the electrode membrane. Only after suppressing these special interferences was a continuous recording of cyanide content possible in this important field of application. The apparatus described covers a concentration range of two orders of magnitude. Due to the lawful limit of 0.1 mg/l CN^-, most operated in the 0.01 to 1.0 mg/l range, although it must be noted that the region below 0.1 mg/l is "asymptotically distorted", since this coincides with the detection limit of the cyanide electrode used. With a 0.1 mg/l cyanide solution the following specifications hold: Drift – about ±2 mV per week; response time – 5 minutes to 90% of final value; accuracy with automatic periodic calibration (once a day) – about ±5%. 5 l of conditioning solution are used a day, so a 50 l container allows for operation for a full week without maintenance. The slope should be checked every 14 days; likewise a rinsing with dilute HCl, in case the sample leads to precipitate formation (for example nickel bath discharge).

The Monitor System Series 1,000 of Orion. Since the detection limit of the AgI/Ag_2S cyanide electrode lies just below 0.1 mg/l, the Orion firm used a different method to indicate cyanide [351]. An Ag_2S electrode is employed, which is known to indicate Ag^+ ions down to $< 10^{-15}$ M. A $KAg(CN)_2$ is used as silver ion buffer to produce a definite response. The silver existing as free ions according to the equilibrium:

$$Ag(CN)_2^- \rightleftharpoons Ag^+ + 2\ CN^- \qquad (R39)$$

is indicated by the electrode. Through the introduction of additional free cyanide (sample solution) the equilibrium (R39) is shifted to the left and the activity of free silver ion is reduced. Since this decreases with the square of the cyanide concentration, a ten-fold increase of cyanide concentration gives rise to an EMF change of −118 mV at the silver electrode, to only −59 mV at the AgI/Ag_2S electrode. The silver electrode also allows a lower detection limit due to the absence of the AgI "silver ion buffer", which obviously leads to higher silver ion concentrations. Since the Orion monitor is not designed specifically for end control of cyanide detoxication systems, the sample handling is different.

The basis of this cyanide monitor is a cell consisting of a silver-selective Ag_2S solid-state membrane electrode and a sodium-selective glass membrane electrode. By substituting a second ion-selective electrode for the usual reference electrode, a cell without liquid junction results (see Chap. 2.3.5) and all of the problems associated with a reference electrode diaphragm (liquid junction potential, clogging, sample contamination) are avoided. On the other hand, the high-ohmic Ag_2S electrode requires a differential amplifier input and due to its unsymmetric construction, the cell gives rise to difficulties mentioned in Chap. 6.2.1.2 concerning the automatic temperature compensation via the isotherm intersection point. For this reason the Orion cyanide monitor possesses an electronically controlled, thermostated measuring cell. In addition to a photo of the instrument, Fig. 63 shows a flow scheme. A four channel proportional pump sucks in the sample (also, if desired, an automatically synchronized calibration standard) and two conditioning solutions. At point *A* the sample is treated with an acetic acid-EDTA solution and then warmed to 80°C in the heat exchanger at point *B* in order to set free all the Ni^{2+} and Cu^+ complex bound cyanide. After removing gas bubbles at *C*, the mixture is redirected to the peristaltic pump and then to point *D* for the final preparation [pH establishment at > pH 11 with 0.1 M NaOH, addition of K [Ag $(CN)_2$]. The addition of NaOH in such large concentrations automatically provides for a stable potential at the "sodium reference electrode"; the addition of acidic EDTA solution hinders the formation of interfering precipitates. The remaining specifications are similar to those of the instrument described above.

6.2.2 Concentration Determination Via a Continuous Titration

If higher accuracies are desired in on-line measurements, titration procedures must again be employed. Figure 64 illustrates the principle involved. On the basis of the EMF coming from the cell, an electronic device automatically adjusts for the maintenance of a pre-determined EMF corresponding to the equivalence point. Usually the sample stream is held constant and the titrant stream is varied according to the EMF difference. In this case the unknown sample concentration can be calculated from the ratio of sample solution to titrant solution flow rates and a knowledge of the titer of the latter. *In contrast to the measurement of a burette volume, which can easily be determined to within 0.1%, the analysis accuracy here depends on the accuracy of the flow rate measurements.*

Continuous titrations can be especially elegantly automated with coulometric reagent production. Here no sophisticated electronic or pneumatomechanic flow regulator is needed to regulate the titrating agent flow rate. The generator current, which is easily controlled with compact electronics (operational amplifiers), is a direct measure of the sample concentration. It is beyond the scope of this book to delve further into the details of such titration processes. Interested readers are referred to the appropriate literature [241].

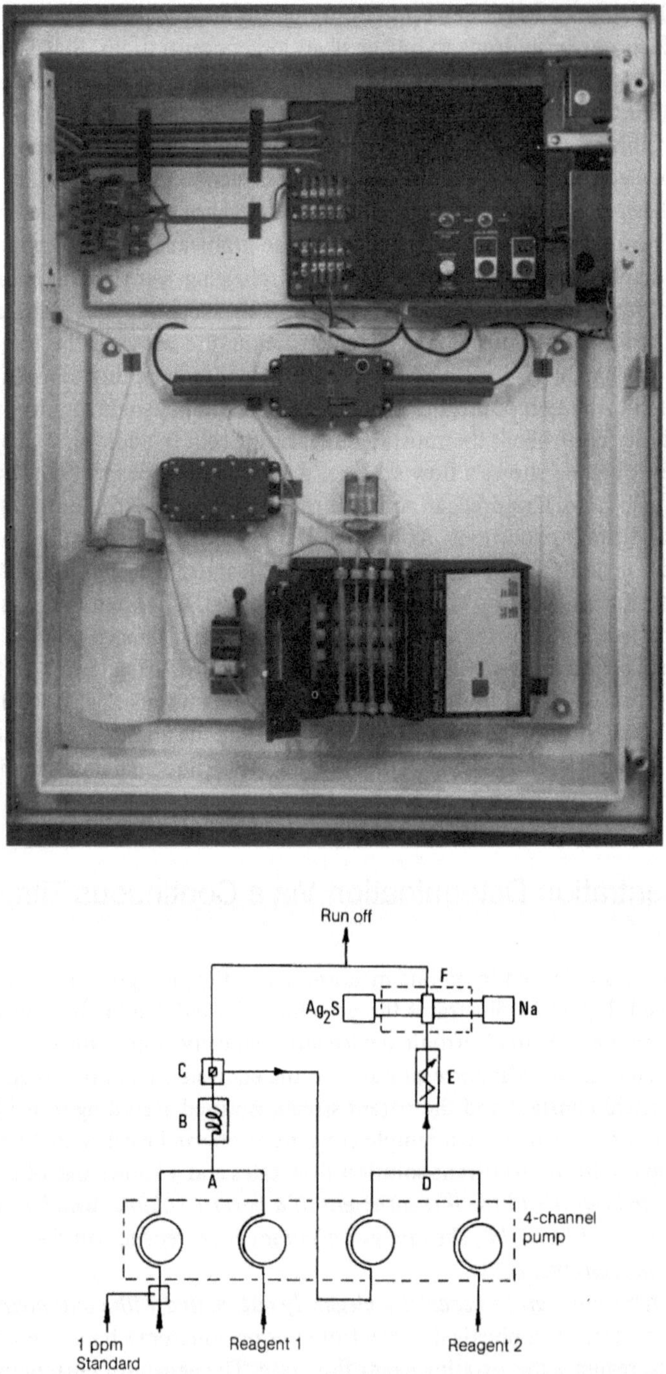

Fig. 63. Cyanide monitor (Orion). Above: Instrument with front plate removed; Below: Flow diagram

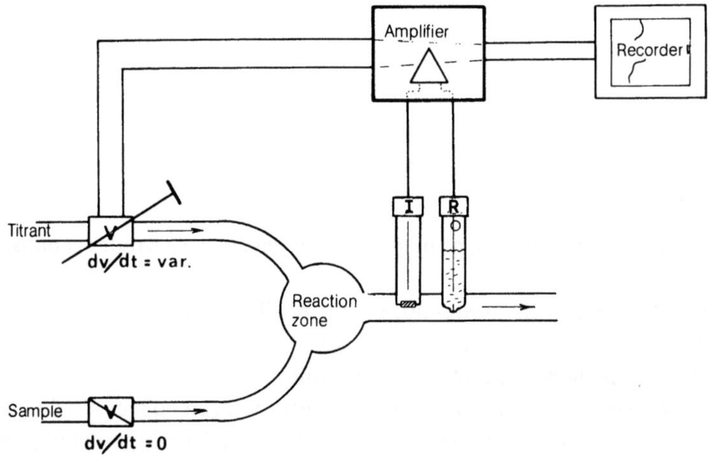

Fig. 64. Continuous titration flow diagram

Outlook

This introduction to the operating principles of ion-selective electrodes is indispensable for:

the self-construction of electrodes not commercially available,

the discovery of new fields of application and

a successful "know-how" during troubleshooting.

It has been shown how the ion activity and concentration can be measured with the help of these electrodes. We have seen that through the choice of a suitable standardisation technique, complex-bound ions can be determined, as well as the ratio of free to bound ions.

Throughout this presentation, the possible sources of error were repeatedly pointed out, since these are usually given too little attention.

Finally, the most important requirements for the successful use of ion selective electrodes are once again enumerated:

Ion-selective electrodes respond to the activity of the indicated ion in solution. With the exception of their use as titration end point indicators, they require calibration with a standard solution of known activity. In the determination of concentrations, the error depends in large part on the validity of the assumed equality and constancy of activity coefficients in the sample and calibration solutions. In order to keep additional error contributions small, the following points should be taken into consideration:

1. The selectivity of the indicating electrode must be sufficient for the desired accuracy.

2. The input impedance of the amplifier used should be at least a factor of 1,000 greater than electrode resistance.

3. The amplifier must be correspondingly linear and guarantee the required read-off precision. Since most direct-indicating instruments (with good ribbon suspension) are good only to about 0.35% and the read-off precision is limited, more accurate work requires an instrument with scale expansion, a compensation device, or preferably a device with digital display.

4. If one wants to increase the accuracy to about 0.001 pa units, then the temperature compensation must also be carried out "properly" via the isotherm intersection point – this is often overlooked. For example, an accuracy of 0.001 pa units over a calibrated range of ±0.5 pa units of the measured ion requires a temperature control over both solutions of 0.6°C and an equally precise setting of the temperature potentiometer. For calibration over a ±2 pa units range this tolerance is reduced to ±0.15°C! The temperature compensator, however, is often only calibrated to ±1°C.

5. The activity of the calibration solution should correspond to that of the sample solution. For accuracies better than 0.01 pa units, the sample measurements should

be followed by a calibration check with two standard solutions, whose measured ion activities bracket those of the samples.

6. As point 4 indicates, a precise thermostating of the measuring cell is recommended for accuracies less than ±0.01 pa unit. Due to the unavoidable uncertainties involved in an "isotherm intersection point compensation" and the fact that only a very few laboratory instruments are equipped for such calibration, this procedure should be avoided by simply making calibration and sample solution measurements at the same temperature.

7. *Cells without liquid junction are to be employed whenever possible.* For measurements with salt bridges, sleeve diaphragms should be chosen for their ability to provide a stable liquid junction potential (with double salt bridge reference electrodes the pH of the outer electrolyte should be adjusted to match that of the sample). In addition, the sample and salt bridge electrolyte solutions should have similar ionic strengths. Cells with liquid junction seldom allow accuracies better than 0.01 pa units.

8. In trace analysis, constant hydrodynamic conditions (stirring, immersion depth and angle, etc.) must be provided for, and streaming potentials must be calibrated for.

Considering these points, the following analysis accuracies are possible with ion-selective electrodes, depending on the interfering ion level[1]:

	Monovalent Ions	Divalent Ions
Concentration determinations with a single standard addition (addition or subtraction techniques, as well as indirect analyses)	1 to 10%	4 to 20%
Activity and concentration determinations using direct potentiometry	1 to 5%	2 to 10%
Concentration determination with the Gran method	0.5 to 3%	
Concentration determination via the evaluation of a titration curve	~ < 0.5%	

This author is of the opinion that the development of ion-selective electrodes is not yet at an end. It is difficult to predict the course of future developments of these sensors, whether selectivity will be attained with semiconductor materials having defined lattice imperfections [440] or through macromolecules with specially designed cavities [136].

The success already achieved with enzyme electrodes leads one to believe that ion-selective microelectrodes will become increasingly important in the future as bases for extremely specific bio-sensors (Bio-Probes, [173, 454]).

Semiconductor technology will also prove to play an important role in the future of ion-selective electrodes as shown by the first attempts using ion-selective transistor electrodes [441–443]. In this case the metal gate of a MOS-FET unit was replaced with a thin (> 100 μ) ion-selective membrane, contacting the sample solution. With

[1] For a detailed survey of all sources of errors involved in this analytical method see [460].

a potassium-selective valinomycin-PVC membrane, the current proportional to the logarithm of the potassium ion activity always lay above 10 μA, and was detectable without further amplification with an inexpensive moving coil instrument [445] (see Fig. 65). Although a few practical problems still remain to be solved (hindering the

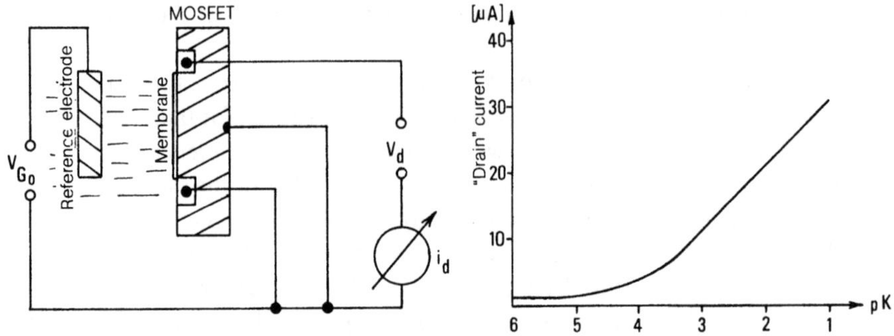

Fig. 65. Ion-selective transistor electrode. Left: Circuit; Right: Calibration curve

dissolution of valinomycin and plasticizer, producing more resistant active transistor surfaces by coating with Si_3N_4, etc.), this surprising success is a strong impulse in the direction of a less expensive technology. In addition, it provides some new and interesting aspects for the as yet unsettled theory of ion-selective electrodes, since according to the circuit (Fig. 65) no current flows across the interior ion-selective membrane/transistor gate interface!

From the point of view of applications, the use of these electrodes as sensitive end point indicators will continue to increase. Surely the possibilities for continuous recording of particular ionic species will become more widely realized, especially in environmental control detectors.

For further applications of ion-selective electrodes refer to the literature citations of the manufacturer. More than 1,000 citations can already be found there.

In the end, the further applications of this analytical technique, which can be thought of as a renaissance in potentiometry, will depend on the skill and experimental innovation of the user. To this end the author hopes to have made some small contribution.

Appendix

A.1 Concentration, Activity and Activity Coefficient

In the previous sections, the activity of the measured ion in solution was often discussed. It was also mentioned, that in contrast to other techniques, ion-selective electrodes respond to the activity rather than the concentration of the free indicated ion characteristic for the electrode under consideration. It was further pointed out that such behavior is in no way a disadvantage, since for many scientific purposes the *effective working concentration,* represented by the activity, is much more important than the actual concentration. If one is interested in the absolute concentration, there are a number of techniques with which the concentration can be very easily obtained. Thus one has the *free choice* between measuring directly in either concentration or activity units, whereas converting the first into the other requires more information (ionic strength) about the sample solution.

Since experience has shown that confusion sometimes arises in the interpretation of electrochemical data, the distinction between activity and concentration, as well as between individual and mean activity coefficients will be elaborated on below [444]:

For dissolved molecules in solution, one can distinguish between:

a) *non-electrolytes* (many organic substances, such as sugar), whose molecules come close enough so that the interaction forces between the dissolved species become effective only at very high concentrations, and

b) *electrolytes,* which dissociate more or less completely into charged ions (mostly inorganic acids, bases and salts). In this case the interactions between particles becomes important at much lower concentrations (~0.01 M), due to the far-reaching electrostatic forces of the ionic charges.

In this sense a solution of a non-electrolyte, arising from the solvation of uncharged molecules, is called an ideal solution. A solution is ideal as long as the chemical potential can be expressed by:

$$\mu_{i(\mathrm{ideal})} = \mu_i^0 + R \cdot T \cdot \ln x_i \tag{A1}$$

where x_i is in concentration units, such as mole fraction. In all other cases one refers to a real solution and introduces a special correction factor f_i for each dissolved species to take into account interactions between solute molecules. This factor must be multiplied by the actual concentration in order for the above equation to hold for real systems as well:

$$\mu_{i(\mathrm{real})} = \mu_i^0 + R \cdot T \cdot \ln x_i \cdot f_i. \tag{A2}$$

The phenomenologically effective, thus thermodynamically relevant, concentration $f_i \cdot x_i$ is called the "activity", and the numerical factor f_i the "individual activity coefficient". This can be only approximately calculated, since the difference in chemical potential $(\mu_{i(real)} - \mu_{i(ideal)})$ represented by the expression $R \cdot T \cdot \ln f_i$ is equal to the electrostatic portion of the work. This electrostatic influence on the chemical potential can be calculated according to a theory of Debye and Hückel, which assumes a point charge surrounded by a uniform cloud of counter ions. As a first approximation, one obtains with this treatment:

$$\mu_{i(real)} - \mu_{i(ideal)} = R \cdot T \cdot \ln f_i = \frac{N_A \, (z_i \cdot e_0)^2}{2 \cdot \epsilon \cdot \kappa^{-1}} \tag{A3}$$

where: N_A = Avogadro's number = 6.023×10^{23} atoms, molecules/mol

e_0 = elementary charge = $1.602 \times 10^{-19}\ A \cdot s$

ϵ = dielectric constant of the medium

κ^{-1} = radius of the ionic cloud, or Debye-Hückel reciprocal length, calculated from:

$$\kappa^{-1} = \sqrt{\frac{\epsilon \cdot k \cdot T}{4\pi} \cdot \frac{1}{\sum_i n_i^0 (z_i \cdot e_0)^2}} \tag{A4}$$

where: k = Boltzmann constant = 1.38×10^{-16} erg/K

n_i^0 = concentration of i.

The theoretically approximate individual activity coefficients f_i defined with Equations (A2) and (A3) are, however, only of theoretical interest, since they cannot be experimentally verified. By definition, only the interactions between ions of a particular type i are expressed, so only this ion can be present in solution for f_i to be experimentally determined. Even if this were succesfully accomplished by means of some fancy apparatus, such as a mass spectrometer, the solution would become increasingly more charged with the introduction of each new i ion. According to Definition (A2), however, a charged solution cannot be present since one starts from the chemical potential μ, and not from the electrochemical potential $\bar{\mu}$ (see Chap. 1). This problem can only be overcome by introducing a neutral, i.e. overall uncharged salt into the solution. In this case one must be satisfied with measuring only the "mean activity coefficients" $f_\pm$ of the cation and the anion. Separation of this change in free energy due to ion-ion interactions into the free energy of cation and anion portions is not experimentally possible. Although this may appear to be the case with pH measurements, more careful investigation will show that a convention was agreed upon. The same is true for pIon values. It would be quite desirable if the U.S. Bureau of Standards were as efficient in this area as in the pH standard area. The first indications of this already exist [445, 446]. Precisely this problem leads to difficulties in interpreting electrochemical data, as well as in preparing solutions of known activity.

A.1.1 Preparing Accurate Activity Calibration Solutions

Since this topic lies at the basis of all measurements with ion-selective electrodes and it is discussed too little in the literature, a pragmatic introduction to these mean activity coefficients will be supplied in some detail below.

Let us start with a simple example and consider the chemical potential of a 1:1 valent electrolyte such as KCl. This is composed of the chemical potentials of the anions and cations. If we relate the mol ratios with x_{K^+} and x_{Cl^-}, then we obtain:

$$\mu_{K^+} = \mu^o_{K^+} + R \cdot T \cdot \ln x_{K^+} + R \cdot T \cdot \ln f_{K^+} \tag{A5}$$

$$\mu_{Cl^-} = \mu^o_{Cl^-} + R \cdot T \cdot \ln x_{Cl^-} + R \cdot T \cdot \ln f_{Cl^-} \tag{A6}$$

$$\mu_{K^+} + \mu_{Cl^-} = \mu^o_{K^+} + \mu^o_{Cl^-} + R \cdot T \cdot \ln x_{K^+} \cdot x_{Cl^-} + R \cdot T \cdot \ln f_{K^+} \cdot f_{Cl^-} . \tag{A7}$$

This sum corresponds to the work done in transferring 2 moles of ions (one mole of K^+ and one mole of Cl^-) from infinity into the solution. In order to refer to the chemical potential of 1 mole of transferred ions, one is talking about a mean chemical potential and mean activity coefficients $f_\pm$, and Eq. (A7) must be divided by two:

$$\frac{\mu_{K^+} + \mu_{Cl^-}}{2} = \frac{\mu^o_{K^+} + \mu^o_{Cl^-}}{2} + \frac{R \cdot T \cdot \ln x_{K^+} \cdot x_{Cl^-}}{2} + \frac{R \cdot T \cdot \ln f_{K^+} \cdot f_{Cl^-}}{2} \tag{A8}$$

where: $$\frac{\mu_{K^+} + \mu_{Cl^-}}{2} = \mu_\pm \tag{A9}$$

$$\frac{\mu^o_{K^+} + \mu^o_{Cl^-}}{2} = \mu^o_\pm \tag{A10}$$

$$\frac{R \cdot T \cdot \ln x_{K^+} \cdot x_{Cl^-}}{2} = R \cdot T \cdot \ln (x_{K^+} \cdot x_{Cl^-})^{1/2} \tag{A11}$$

$$\frac{R \cdot T \cdot \ln f_{K^+} \cdot f_{Cl^-}}{2} = R \cdot T \cdot \ln (f_{K^+} \cdot f_{Cl^-})^{1/2} \tag{A12}$$

$$(x_{K^+} \cdot x_{Cl^-})^{1/2} = x_\pm \tag{A13}$$

$$(f_{K^+} \cdot f_{Cl^-})^{1/2} = f_\pm . \tag{A14}$$

Finally we obtain the experimentally accessible quantity:

$$\mu_\pm = \mu^o_\pm + R \cdot T \cdot \ln x_\pm + R \cdot T \cdot \ln f_\pm . \tag{A15}$$

For electrolytes which dissociate into υ^+ cations and υ^- anions, in general:

$$f_\pm = (f_+^{\upsilon^+} \cdot f_-^{\upsilon^-})^{1/\upsilon} \quad \text{(A16)}$$

where:
$$\upsilon = \upsilon_+ + \upsilon_- , \quad \text{(A17)}$$

or taking the logarithm:

$$\ln f_\pm = \frac{1}{\upsilon}(\upsilon_+ \cdot f_+ + \upsilon_- \cdot f_-). \quad \text{(A18)}$$

If this equation is substituted into the theoretically derived Expression (A3), then

$$\ln f_\pm = -\frac{1}{\upsilon}\left[\frac{N_A \cdot e_0^2 \cdot \kappa}{2 \cdot \epsilon \cdot R \cdot T}(\upsilon_+ z_+^2 + \upsilon_- z_-^2)\right]. \quad \text{(A19)}$$

Rearrangement, conversion to mole/l concentration units and collecting the constants in the Debye-Hückel approximation for a very dilute electrolyte (< 0.01 M/l) yields:

$$\log f_\pm = -A(z_+ \cdot z_-) \cdot I^{1/2} \quad \text{(A20)}$$

where: A = medium- and temperature-dependent constant (for water at 25°C A = 0.512)

z_+ = valence of cation

z_- = valence of anion

I = total ionic strength of the solution, defined as $I = \frac{1}{2} \sum_i c_i \cdot z_i^2$ where c_i = concentration of ion i in mol/l.

Example: For a 0.1 M KCl solution:

$$I = \frac{1}{2}(0.1 \times 1^2 + 0.1 \times 1^2) = 0.1. \quad \text{(A21)}$$

For a 0.1 M $CaCl_2$ solution:

$$I = \frac{1}{2}(0.1 \times 2^2 + 0.2 \times 1^2) = 0.3. \quad \text{(A22)}$$

Equation (A20) indicates that the activity coefficient is independent of the particular ionic species. It depends only on the type of electrolyte (cation valence, anion valence, concentration).

Experimentally, one finds deviations which exhibit individual character above 0.01 M electrolyte concentration. The individual characteristic neglected in the deviation of Equation (A3) is the effective size of the solvated ion, which has a characteristic value for each particular kind of ion. Introducing a radius a, which can approximate that of the ion pair, we obtain:

$$\log f_{\pm} \approx -\frac{A \cdot z_i^2 \cdot I^{1/2}}{1 + B \cdot a \cdot I^{1/2}} \tag{A23}$$

where A, B = medium- and temperature-dependent constants.

For water at 25°C, for example $A = 0.512$
$B = 0.329 \times 10^8$.

As this equation shows, in order to convert from concentration to activity units, not only must the total ionic strength be known, but also the approximate effective radius a. Since this can only be determined to a certain accuracy, the activity cannot be calculated any more accurately. This does not mean that in general activities can only be determined approximately with ion-selective electrodes. On the contrary, electrodes are calibrated with very dilute solutions ($< 10^{-4}$ M), so the activity coefficients need not be taken into account, since within the limits of the measurement accuracy they can be set equal to 1. Only if high activities obtained with such a calibration are

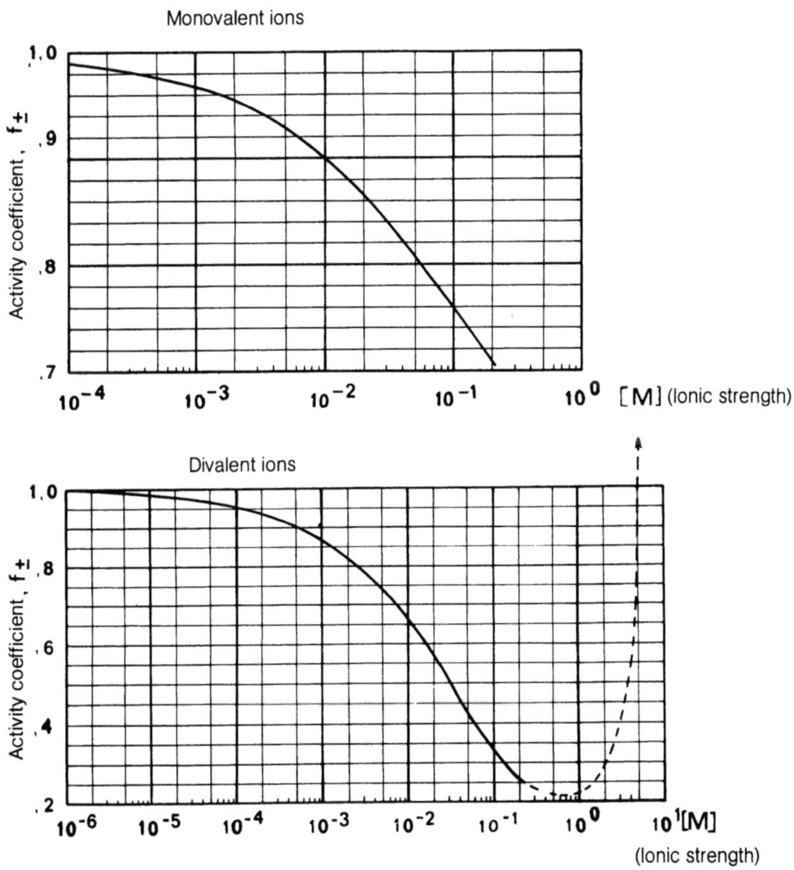

Diagram A1

to be converted into the corresponding concentrations does the above problem arise. In this case one is better off using a different technique which directly indicates concentration.

Diagram A1 shows the dependence of the mean activity coefficient on ionic strength. The coefficients calculated according to the theory of interionic interactions discussed in the previous section show excellent agreement with those obtained experimentally with ionic strengths below 1 M. This theory cannot explain the rapidly increasing mean activity coefficients measured at higher ionic strengths. In a purely qualitative fashion, this increase in effective concentration, thus the activity, can be explained by considering the fact that with continually increasing concentration of electrolyte, the concentration of free solvent molecules continually decreases. The solvent molecules bound in the solvation sphere of the ion should not be considered as free solvent molecules [447]. These molecules do not perform any more work during further dissolution processes. Thus a lowering of the effective solvent concentration results, which would be reflected in an increasing average activity coefficient. This can be taken into account mathematically by introducing a further correction term: $-0.2 \cdot I$ (water, 25°C).

A.2 Survey of the Temperature Dependence of Commonly Used Reference Electrodes

Standard Potential: ($\epsilon^0 + \epsilon_j$) [V]

T [°C]	Ag/AgCl [54] 3.5 M KCl	Ag/AgCl [54] sat. KCl	Thalamide [62] sat. KCl	Calomel [54] 3.5 M KCl	Calomel [54] sat. KCl
5			−0.5645		
10	0.2152	0.2138	−0.5684	0.2556	0.2542
15	0.2117	0.2089	−0.5720	0.2538	0.2510
20	0.2082	0.2040	−0.5755	0.2520	0.2478
25	0.2046	0.1989	−0.5794	0.2501	0.2444
30	0.2009	0.1939	−0.5837	0.2481	0.2411
35	0.1971	0.1887	−0.5878	0.2460	0.2376
40	0.1933	0.1835	−0.5915	0.2439	0.2341
50			−0.6000		
60			−0.6084		
70			−0.6164		
80			−0.6242		
90			−0.6311		

Electrode correction relative to the normal hydrogen electrode:

$$E = \frac{2.3026\, R \cdot T}{2F} \log \frac{760}{p_{H_2}^*}$$

$$p_{H_2}^* = b - p_{H_2O} + \frac{h}{34}$$

where: b = barometer reading in [torr]

p_{H_2O} = partial pressure of water at the measuring temperature in [torr]

h = average immersion depth of the H_2 inlet tube (electrode depth for autogeneration of H_2) in [mm].

Numerical values of the Nernst constant ($\frac{R \cdot T}{F}$) at various temperatures.

T [°C]	$K_t = \frac{mV}{pa}$	T [°C]	$K_t = \frac{mV}{pa}$
0	54.20	50	64.12
5	55.19	55	65.11
10	56.18	60	66.10
15	57.17	65	67.09
20	58.17	70	68.09
25	59.16	80	70.07
30	60.15	90	72.05
35	61.14	100	74.04
40	62.13	135	81.00
45	63.13		

A.3 Table for Evaluation Using Analysis Technique 5.4.1

Measurement of the EMF change upon addition of a standard solution to the sample solution (according to Orion).

The table assumes that measurements are carried out at 25°C, that 100 ml of sample solution are used and that 10 ml of a 10-fold more concentrated standard solution are added. The theoretical electrode slope is assumed. For cations the ΔE values should have the opposite sign and for divalent ions the ΔE values should be doubled. The concentration of measured ion in the sample solution is obtained by multiplying the appropriate standard solution concentration by the factor f corresponding to the measured ΔE.

ΔE [mV]	f	ΔE [mV]	f	ΔE [mV]	f	ΔE [mV]	f
		−3.0	0.423	−8.5	0.188	−18.0	0.0822
+2.4	52.6	−3.1	0.415	−8.6	0.186	−18.2	0.0811
+2.3	17.2	−3.2	0.407	−8.7	0.184	−18.4	0.0799
+2.2	10.3	−3.3	0.399	−8.8	0.182	−18.6	0.0788
+2.1	7.32	−3.4	0.391	−8.9	0.180	−18.8	0.0777
+2.0	5.68	−3.5	0.384	−9.0	0.178	−19.0	0.0767
+1.9	4.63	−3.6	0.377	−9.1	0.176	−19.2	0.0756
+1.8	3.91	−3.7	0.370	−9.2	0.174	−19.4	0.0746
+1.7	3.38	−3.8	0.363	−9.3	0.173	−19.6	0.0736
+1.6	2.98	−3.9	0.357	−9.4	0.171	−19.8	0.0726
+1.5	2.66	−4.0	0.351	−9.5	0.169	−20.0	0.0716
+1.4	2.40	−4.1	0.345	−9.6	0.167	−20.2	0.0707
+1.3	2.19	−4.2	0.339	−9.7	0.165	−20.4	0.0698
+1.2	2.01	−4.3	0.333	−9.8	0.164	−20.6	0.0689
+1.1	1.86	−4.4	0.327	−9.9	0.162	−20.8	0.0680
+1.0	1.72	−4.5	0.322	−10.0	0.160	−21.0	0.0671
+0.9	1.61	−4.6	0.319	−10.2	0.157	−21.2	0.0662
+0.8	1.51	−4.7	0.312	−10.4	0.154	−21.4	0.0654
+0.7	1.42	−4.8	0.307	−10.6	0.151	−21.6	0.0645
+0.6	1.34	−4.9	0.302	−10.8	0.148	−21.8	0.0637
+0.5	1.27	−5.0	0.297	−11.0	0.145	−22.0	0.0629
+0.4	1.21	−5.1	0.293	−11.2	0.143	−22.2	0.0621
+0.3	1.15	−5.2	0.288	−11.4	0.140	−22.4	0.0613
+0.2	1.09	−5.3	0.284	−11.6	0.137	−22.6	0.0606
+0.1	1.05	−5.4	0.280	−11.8	0.135	−22.8	0.0598
0.0	1.00	−5.5	0.276	−12.0	0.133	−23.0	0.0591
−0.1	0.959	−5.6	0.272	−12.2	0.130	−23.2	0.0584
−0.2	0.921	−5.7	0.268	−12.4	0.128	−23.4	0.0576
−0.3	0.886	−5.8	0.264	−12.6	0.126	−23.6	0.0569
−0.4	0.853	−5.9	0.260	−12.8	0.123	−23.8	0.0563

ΔE [mV]	f	ΔE [mV]	f	ΔE [mV]	f	ΔE [mV]	f
−0.5	0.822	−6.0	0.257	−13.0	0.121	−24.0	0.0556
−0.6	0.794	−6.1	0.253	−13.2	0.119	−24.2	0.0549
−0.7	0.767	−6.2	0.250	−13.4	0.117	−24.4	0.0543
−0.8	0.742	−6.3	0.247	−13.6	0.115	−24.6	0.0536
−0.9	0.718	−6.4	0.243	−13.8	0.113	−24.8	0.0530
−1.0	0.696	−6.5	0.240	−14.0	0.112	−25.0	0.0523
−1.1	0.675	−6.6	0.237	−14.2	0.110	−25.2	0.0517
−1.2	0.655	−6.7	0.234	−14.4	0.108	−25.4	0.0511
−1.3	0.637	−6.8	0.231	−14.6	0.106	−25.6	0.0505
−1.4	0.619	−6.9	0.228	−14.8	0.105	−25.8	0.0499
−1.5	0.602	−7.0	0.225	−15.0	0.103	−26.0	0.0494
−1.6	0.586	−7.1	0.222	−15.2	0.1013	−26.2	0.0488
−1.7	0.571	−7.2	0.219	−15.4	0.0997	−26.4	0.0482
−1.8	0.556	−7.3	0.217	−15.6	0.0982	−26.6	0.0477
−1.9	0.542	−7.4	0.214	−15.8	0.0967	−26.8	0.0471
−2.0	0.529	−7.5	0.212	−16.0	0.0952	−27.0	0.0466
−2.1	0.516	−7.6	0.209	−16.2	0.0938	−27.2	0.0461
−2.2	0.504	−7.7	0.207	−16.4	0.0924	−27.4	0.0456
−2.3	0.493	−7.8	0.204	−16.6	0.0910	−27.6	0.0450
−2.4	0.482	−7.9	0.202	−16.8	0.0897	−27.8	0.0445
−2.5	0.471	−8.0	0.199	−17.0	0.0884	−28.0	0.0440
−2.6	0.461	−8.1	0.197	−17.2	0.0871	−28.2	0.0435
−2.7	0.451	−8.2	0.195	−17.4	0.0858	−28.4	0.0431
−2.8	0.441	−8.3	0.193	−17.6	0.0846	−28.6	0.0426
−2.9	0.432	−8.4	0.190	−17.8	0.0834	−28.8	0.0421

A.4 Table for Evaluation Using Analysis Technique 5.4.2

Measurement of the EMF change upon addition of the sample solution to a standard solution (according to Orion).

The table assumes that measurements are carried out at 25°C and that the electrochemical cell exhibits the theoretical Nernst slope. The concentration of the measured ion in the sample solution is obtained by multiplying the expression $\frac{V_s \cdot C_s}{V_t}$ by the factor f corresponding to the measured ΔE.

Here V_s = starting volume of the standard solution

C_s = concentration of the standard solution

V_t = volume of the added test (sample) solution.

ΔE [mV]		Addition method	Subtraction method
monovalent ions	divalent ions	f	f
1.0	0.5	0.040	0.038
1.2	0.6	0.048	0.046
1.4	0.7	0.056	0.053
1.6	0.8	0.064	0.060
1.8	0.9	0.073	0.068
2.0	1.0	0.081	0.075
3.0	1.5	0.12	0.11
4.0	2.0	0.17	0.14
5.0	2.5	0.21	0.18
6.0	3.0	0.26	0.21
7.0	3.5	0.31	0.24
8.0	4.0	0.36	0.27
9.0	4.5	0.42	0.30
10.0	5.0	0.48	0.32
11.0	5.5	0.53	0.35
12.0	6.0	0.60	0.37
13.0	6.5	0.66	0.40
14.0	7.0	0.72	0.42
15.0	7.5	0.79	0.44
16.0	8.0	0.86	0.46
17.0	8.5	0.94	0.48
18.0	9.0	1,01	0.50
19.0	9.5	1.09	0.52
20.0	10.0	1.18	0.54

ΔE [mV]		Addition method	Subtraction method
monovalent ions	divalent ions	f	f
21.0	10.5	1.26	0.56
22.0	11.0	1.35	0.58
23.0	11.5	1.45	0.59
24.0	12.0	1.55	0.61
25.0	12.5	1.65	0.62
26.0	13.0	1.75	0.64
27.0	13.5	1.86	0.65
28.0	14.0	1.97	0.66
29.0	14.5	2.09	0.68
30.0	15.0	2.21	0.69
31.0	15.5	2.34	0.70
32.0	16.0	2.47	0.71
33.0	16.5	2.61	0.72
34.0	17.0	2.76	0.73
35.0	17.5	2.90	0.74
36.0	18.0	3.06	0.75
37.0	18.5	3.22	0.76
38.0	19.0	3.39	0.77
39.0	19.5	3.56	0.78
40.0	20.0	3.74	0.79

A.5 Table for Evaluation Using Analysis Technique 5.5.1

Method of double standard addition (according to Orion).
For the meaning of the symbols see Sect. 5.5.1.

R	$Cx/\Delta C$	R	$Cx/\Delta C$	R	$Cx/\Delta C$	R	$Cx/\Delta C$
1.270	0.100	1.495	0.600	1.620	1.213	1.720	2.126
1.280	0.113	1.500	0.618	1.625	1.245	1.725	2.190
1.290	0.126	1.505	0.637	1.630	1.280	1.730	2.256
1.300	0.140	1.510	0.655	1.635	1.315	1.735	2.326
1.310	0.154	1.515	0.675	1.640	1.353	1.740	2.397
1.320	0.170	1.520	0.694	1.645	1.391	1.745	2.470
1.330	0.186	1.525	0.714	1.650	1.430	1.750	2.549
1.340	0.203	1.530	0.735	1.655	1.469	1.755	2.629
1.350	0.221	1.535	0.756	1.660	1.510	1.760	2.711
1.360	0.240	1.540	0.778	1.665	1.554	1.765	2.801
1.370	0.260	1.545	0.801	1.670	1.598	1.770	2.892
1.380	0.280	1.550	0.823	1.675	1.643	1.775	2.985
1.390	0.302	1.555	0.847	1.680	1.691	1.780	3.088
1.400	0.325	1.560	0.870	1.685	1.738	1.785	3.193
1.410	0.349	1.565	0.896	1.690	1.787	1.790	3.301
1.420	0.373	1.570	0.920	1.695	1.840	1.795	3.416
1.430	0.399	1.575	0.946	1.700	1.894	1.800	3.536
1.440	0.427	1.580	0.973	1.705	1.948	1.805	3.664
1.450	0.455	1.585	1.000	1.710	2.006	1.810	3.797
1.460	0.485	1.590	1.029	1.715	2.066	1.815	3.939
1.470	0.516	1.595	1.056				
1.475	0.532	1.600	1.086				
1.480	0.548	1.605	1.116				
1.485	0.565	1.610	1.147				
1.490	0.582	1.615	1.179				

A.6 Table for Evaluation Using Analysis Technique 5.5.2

Method of standard addition with subsequent dilution (according to Orion).

The table is based on a 1:1 dilution. For an explanation of the symbols see Sect. 5.5.2.

R	*Cx/ΔC*	*R*	*Cx/ΔC*	*R*	*Cx/ΔC*	*R*	*Cx/ΔC*
0.20	6.727	0.80	1.349	1.30	0.684	1.80	0.403
0.22	6.072	0.82	1.307	1.32	0.668	1.82	0.395
0.24	5.526	0.84	1.266	1.34	0.653	1.84	0.388
0.26	5.066	0.86	1.227	1.36	0.638	1.86	0.380
0.28	4.670	0.88	1.190	1.38	0.624	1.88	0.373
0.30	4.327	0.90	1.155	1.40	0.610	1.90	0.366
0.32	4.026	0.92	1.121	1.42	0.597	1.92	0.359
0.34	3.763	0.94	1.089	1.44	0.584	1.94	0.353
0.36	3.528	0.96	1.058	1.46	0.571	1.96	0.346
0.38	3.319	0.98	1.028	1.48	0.559	1.98	0.340
0.40	3.130	1.00	1.000	1.50	0.547	2.00	0.333
0.42	2.959	1.02	0.973	1.52	0.535	2.02	0.327
0.44	2.805	1.04	0.947	1.54	0.524	2.04	0.321
0.46	2.663	1.06	0.922	1.56	0.513	2.06	0.316
0.48	2.533	1.08	0.898	1.58	0.503	2.08	0.310
0.50	2.415	1.10	0.875	1.60	0.492	2.10	0.304
0.52	2.305	1.12	0.852	1.62	0.482	2.12	0.299
0.54	2.203	1.14	0.831	1.64	0.473	2.14	0.294
0.56	2.109	1.16	0.810	1.66	0.463	2.16	0.288
0.58	2.021	1.18	0.790	1.68	0.454	2.18	0.283
0.60	1.939	1.20	0.771	1.70	0.445	2.20	0.278
0.62	1.863	1.22	0.752	1.72	0.436	2.22	0.273
0.64	1.791	1.24	0.734	1.74	0.427	2.24	0.269
0.66	1.724	1.26	0.717	1.76	0.419	2.26	0.264
0.68	1.661	1.28	0.700	1.78	0.411	2.28	0.259
0.70	1.602						
0.72	1.545						
0.74	1.492						
0.76	1.442						
0.78	1.394						

A.7 Evaluation Table for Standard Addition + 1:1 Dilution

(according to Metrohm)

Increment factor as a function of the slope at 25°C (monovalent ions).

ΔE_{ad} [mV]	% theoretical slope										
	100	99	98	97	96	95	94	93	92	91	90
	ΔE_V [mV]										
	17.8	17.6	17.5	17.3	17.1	16.9	16.7	16.6	16.4	16.2	16.0
10	2.10	2.08	2.05	2.03	2.00	1.97	1.95	1.92	1.90	1.87	1.85
11	1.87	1.85	1.83	1.80	1.78	1.70	1.73	1.71	1.69	1.66	1.64
12	1.68	1.66	1.64	1.62	1.60	1.57	1.55	1.53	1.51	1.49	1.47
13	1.52	1.50	1.48	1.46	1.44	1.42	1.40	1.38	1.36	1.34	1.33
14	1.38	1.36	1.34	1.33	1.31	1.29	1.27	1.26	1.24	1.22	1.20
15	1.26	1.24	1.23	1.21	1.19	1.18	1.16	1.14	1.13	1.11	1.10
16	1.16	1.14	1.13	1.11	1.10	1.08	1.06	1.05	1.03	1.02	1.00
17	1.07	1.05	1.04	1.02	1.01	0.99	0.98	0.96	0.95	0.94	0.92
18	0.99	0.97	0.96	0.94	0.93	0.92	0.90	0.89	0.88	0.86	0.85
19	0.91	0.90	0.89	0.87	0.86	0.85	0.84	0.82	0.81	0.80	0.78
20	0.85	0.84	0.82	0.81	0.80	0.79	0.78	0.76	0.75	0.74	0.73
21	0.79	0.78	0.77	0.76	0.74	0.73	0.72	0.71	0.70	0.69	0.68
22	0.74	0.73	0.72	0.71	0.69	0.68	0.67	0.66	0.65	0.64	0.63
23	0.69	0.68	0.67	0.66	0.65	0.64	0.63	0.62	0.61	0.60	0.59
24	0.65	0.64	0.63	0.62	0.61	0.60	0.59	0.58	0.57	0.56	0.55
25	0.61	0.60	0.59	0.58	0.57	0.56	0.55	0.54	0.53	0.52	0.51
26	0.57	0.56	0.55	0.54	0.53	0.53	0.52	0.51	0.50	0.49	0.48
27	0.54	0.53	0.52	0.51	0.50	0.49	0.49	0.48	0.47	0.46	0.45
28	0.51	0.50	0.49	0.48	0.47	0.47	0.46	0.45	0.44	0.43	0.42
29	0.48	0.47	0.46	0.45	0.45	0.44	0.43	0.42	0.41	0.41	0.40
30	0.45	0.44	0.44	0.43	0.42	0.41	0.41	0.40	0.39	0.38	0.38
31	0.43	0.42	0.41	0.40	0.40	0.39	0.38	0.38	0.37	0.36	0.35
32	0.40	0.40	0.39	0.38	0.38	0.37	0.36	0.36	0.35	0.34	0.33
33	0.38	0.38	0.37	0.36	0.36	0.35	0.34	0.34	0.33	0.32	0.32
34	0.36	0.36	0.35	0.34	0.34	0.33	0.32	0.32	0.31	0.30	0.30
35	0.34	0.34	0.33	0.33	0.32	0.31	0.31	0.30	0.29	0.29	0.28
36	0.33	0.32	0.31	0.31	0.30	0.30	0.29	0.28	0.28	0.27	0.27
37	0.31	0.30	0.30	0.29	0.29	0.28	0.28	0.27	0.26	0.26	0.25
38	0.30	0.29	0.28	0.28	0.27	0.27	0.26	0.26	0.25	0.25	0.24
39	0.28	0.28	0.27	0.26	0.26	0.25	0.25	0.24	0.24	0.23	0.23
40	0.27	0.26	0.26	0.25	0.25	0.24	0.24	0.23	0.23	0.22	0.22
41	0.25	0.25	0.24	0.24	0.23	0.23	0.22	0.22	0.21	0.21	0.20
42	0.24	0.24	0.23	0.23	0.22	0.22	0.21	0.21	0.20	0.20	0.19
43	0.23	0.23	0.22	0.22	0.21	0.21	0.20	0.20	0.19	0.19	0.18
44	0.22	0.22	0.21	0.21	0.20	0.20	0.19	0.19	0.18	0.18	0.18
45	0.21	0.21	0.20	0.20	0.19	0.19	0.18	0.18	0.18	0.17	0.17
46	0.20	0.20	0.19	0.19	0.18	0.18	0.17	0.17	0.17	0.16	0.16
47	0.19	0.19	0.18	0.18	0.17	0.17	0.17	0.16	0.16	0.15	0.15
48	0.18	0.18	0.17	0.17	0.17	0.16	0.16	0.15	0.15	0.15	0.14
49	0.17	0.17	0.17	0.16	0.16	0.16	0.15	0.15	0.14	0.14	0.14
50	0.17	0.16	0.16	0.16	0.15	0.15	0.14	0.14	0.14	0.13	0.13

A.8 Ion-Selective Electrode Manufacturers and/or Dealers

Not all commercially available electrodes could be mentioned in this text book because new developments are constantly appearing and old constructions suddenly stopped being produced. In addition it is difficult to differentiate between dealers and manufacturers because of OEM contracts. The following list, even though incomplete, should help the potential user to find a supplier in his area.

Amel, Via Bolzano 30, 20127 Milano (Italy)
Applied Medical Technology Inc., P.O.Box 689, Menlo Park, Calif. 94025 (U.S.A.)
Beckman Instruments Inc., 2500 Harbor Blvd., Fullerton, Calif. 92634 (U.S.A.)
Brinkmann Instruments Inc., (Metrohm), Cantiague Road, Westbury, N.Y. 11590 (U.S.A.)
Chemlab Instruments Ltd., 129 Upminster Road, Hornchurch, Essex RM11 3XJ (England)
Chemtrix Inc., 163 S.W. Freeman Ave., Hillsboro, OR 97123 (U.S.A.)
Corning Medical, 11 North Street, Medfield, Mass. 02053 (U.S.A.)
Electrofact B.V., P.O.Box 163, 3800 AD Amersfoort (Netherlands)
Electronic Instruments Ltd., Hanesworth Lane, Chertsey, Surrey KT 16 9LF (England)
Extech International Corporation, 177 State Street, Boston, Mass. 02109 (U.S.A.)
Forschungsinstitut Meinsberg, 7305 Waldheim PSF73 (DDR)
Graphic Controls, 189 Van Rensselaer Street, Buffalo, N.Y. 14240 (U.S.A.)
HNU Systems Inc., 30 Ossipee Road, Newton, MA 02164 (U.S.A.)
Dr. W. Ingold KG, Brönnerstr. 5-9, 6 Frankfurt (W.Germany)
International Crystal Lab., 107 Trumbull Street, Elizabeth, N.J. 07206 (U.S.A.)
Ionel, RR1 Box 296, Mount Hope, Ontario LOR 1WO (Canada)
Kovo (Crytur), CS-100 00 Praha 10 Strasice (Czechoslovakia)
Lazar Research Labs Inc., 731 La Brea Ave., Los Angeles, Calif. 90036 (U.S.A.)
The London Company (Radiometer), 811 Sharon Drive, Cleveland, OH 44145 (U.S.A.)
Markson Science Inc., Box 767, Del Mar, Calif. 92014 (U.S.A.)
Metrimplex (Radelkis), P.O.Box 202, H-1391 Budapest 62 (Hungary)
Metrohm AG, CH-9100 Herisau (Switzerland)
Microelectrodes Inc., Londonderry, NH 03053 (U.S.A.)
Nova Biochemical, 1238 Chestnut Street, Newton, Mass. 02164 (U.S.A.)
Philips, Eindhoven (Netherlands)
Photovolt Corporation, 1155 Broadway, New York, N.Y. 10010 (U.S.A.)
Polymetron AG, CH-8634 Hombrechtikon ZH (Switzerland)
Pope Scientific Co, P.O.Box 495, Menomee Falls, WI 53051 (U.S.A.)
Presto-Tek Corporation, Figueroa Street, Los Angeles, Calif. 90041 (U.S.A.)
Radiometer A/S, Emdrupvej 72, DK-2400, Copenhagen NV (Denmark)
Sargent-Welch Scientific Co., 7300 North Linden Ave., Skokie, IL 60076 (U.S.A.)
Schott & Gen., Hattenbergstr. 10, 65 Mainz (W.Germany)
Sensorex, 9713 Bosla Ave., Westminster, Calif. 92683 (U.S.A.)
Solea-Tacussel, 72-78 rue d'Alsace, Lyon, F-69100 Villeurbanne (France)
Technicon Industrial Systems, 511 Benedict Ave., Tarrytown, N.Y. 01591 (U.S.A.)
Tudor Scientific Glass Co., 555 Edgefield Road, North Augusta, SC 29841 (U.S.A.)
WP Instruments Inc., 60 Fitch Street, New Haven, CT 06515 (U.S.A.)
Zellweger Uster Inc. (Polymetron), P.O.Box 16348, Charlotte, NC 28216 (U.S.A.)

Literature

1. Gerischer, H.: Z.physik.Chem., N.F. **26**, 325 (1960); **27**, **48** (1961).
2. Dewald, J.F.: Semiconductor electrodes. In: Semiconductors, ed. N.B. Hannay. New York: Reinhold Publishing Corp. 1964.
3. Gerischer, H., Cammann, K.: Ber. Bunsen Ges. **76**, 385 (1972).
4. Cammann, K., Rechnitz, G.A.: Anal. Chem. **48**, 856 (1976).
5. Dole, M.: The glass electrode, methods, applications and theory. New York: J. Wiley & Sons, 1941.
6. Kratz, L.: Die Glaselektrode und ihre Anwendungen. Frankfurt/Main: D. Steinkopff 1950.
7. Eisenman, G.: Glass electrodes for hydrogen and other cations, principles and practice. New York: M. Dekker, 1967.
8. Cremer, M.: Z. Biol. **47**, 562 (1906).
9. Horovitz, K.: Z. Physik **15**, 369 (1923).
10. Horovitz, K.: Z. physik. Chem. **115**, 424 (1925).
11. Nikolsky, B.P.: Z. physik. Chem. (russ.) **10**, 495 (1937).
12. Nikolsky, B.P., Tohnatschewa, T.A.: Z. physik. Chem. (russ.) **10**, 513 (1937).
13. Schiller, H.: Ann. Physik (4) **75**, 105 (1924).
14. Lengyel, B., Blum, E.: Trans. Faraday Soc. **30**, 461 (1934).
15. Ssokolof, S.J., Passynsky, A.H.: Z. physik. Chem. A **160**, 366 (1932).
16. Eisenman, G., Rudin, D.O., Casby, J.U.: Science **126**, 831 (1957).
17. Schwabe, K.: pH-Meßtechnik. Dresden-Leipzig: Th. Steinkopff, 1963.
18. Schwabe, K., Suschke, H.D.: Angew. Chem. **76**, 39 (1964).
19. Eisenman, G.: Biophys. J. **2**, part 2, 259 (1962).
20. Lengyel, B., Csákvári, B.: : Chem. Ztg. **93**, 763 (1969).
21. Eisenman, G.: The ion-exchange characteristics of the hydrated surface of Na^+ selective glass electrodes. In: glass microelectrodes, eds. M. Lavallée, O.F. Schanne a. N.C. Hébert. New York: J. Wiley 1969.
22. Nikolsky, B.P., Schulz, M.M., Peschechonowa, N.V.: Z. physik. Chem. (russ.) **32**, **19** (1958).
23. Baucke, F.G.K.: Non-Christalline Solids **14**, 13 (1974).
24. Hubbard, D., Black, S., Holley, S., Rynders, G.: J. Res. nat. Bur. Stand. **46**, 168 (1961).
25. Tendeloo, H.J.C., Voorspuij, A.J.Z.: Rec. Trav. chim. Pays-Bas **61**, 531 (1942).
26. Shermer, H.F., Rynders, G.F., Cleek, G.W., Hubbard, D.: J. Res. NBS **52**, 251 (1954).
27. Garrels, R.M., Sato, M., Thompson, M.E., Truesdell, A.H.: Science **135**, 1045 (1962).
28. Truesdell, A.H., Pommer, A.M.: Science **142**, 1292 (1963).
29. Truesdell, A.H., Christ, C.L.: In: Glass electrodes for hydrogen and other cations. New York: M. Dekker 1967.
30. Wagner, C., Traud, W.: Z. Elektrochem. **44**, 391 (1938).
31. Kimball, G., Glassner, A.: J. Chem. Phys. **8**, 815 (1940).
32. Bockris, J.O'M.: „Modern Aspects of Electrochemistry", eds: Bockris, Conway Vol.1. London: Butterworth, 1954.
33. Lakshminarayanaiah, N.: Membrane Electrodes, New York, Academic Press 1976
34. Wikbey, A. : J. Electroanal. Chem. **38**, 429 (1972).
35. Cammann, K.: Dissertation, Universität München 1975.
36. Eisenman, G.: Theory of membrane electrode potentials. In: Ionselective electrodes, ed. R.A. Durst. Washinton-NBS, Spec. Publ. 314, 1969.
37. Koryta, J.: Ion-selective electrodes. Cambridge University Press, Cambridge 1975.

38. Friedman, S.M.: In: The glass electrode, eds. G. Eisenman, R. Bates, G. Mattock and S.M. Friedman. New York: Interscience 1966.
39. Rechnitz, G.A., Kugler, G.C.: Anal. Chem. **39**, 1682 (1967)
40. Rechnitz, G.A.: Analytical studies. In: Ion-selective electrodes, ed. R.A. Durst. Washington: NBS, Spec. Publ. 314, 1969.
41. Pungor, E., Tóth, K., Havas, J.: Acta chim. Acad. Sci. hung. **48**, 17 (1966).
42. Buchanan, E.B., Seago, J.L.: Anal. Chem. **40**, 517 (1968).
43. Stefanac, R., Simon, W.: Mikrochem. J. **12**, 125 (1967).
44. Pedersen, C.J.: J. Amer. chem. Soc. **89**, 7017 (1967).
45. Bush, M.A., Truter, M.R.: J. Chem. Soc. Perkin II, 345 (1972).
46. Nikolsky, B.P., Schulz, Belijustinin: In: Glass electrodes for hydrogen and other cations, ed. G. Eisenman. New York: M. Dekker 1967.
47. Moore, E.W.: Ann. N.Y. Acad. Sci. **148** (1), 93 (1968).
48. Huston, R., Butler, J.N.: Anal. Chem. **41**, 200 (1969).
49. Latimer, W.M.: Oxidation potentials. New York: Prentice-Hall Inc. 1953.
50. Feltham, A.M., Spiro, M.: J. Electroanal. Chem. **35**, 181 (1972).
51. Popoff, S., Kunz, A.H., Snow, R.D.: J. Phys. Chem. **32**, 1056 (1928).
52. Hills, G.J., Ives, D.J.G.: J. Chem. Soc. 305, (1951).
53. Giner, J.: J. electrochem. Soc. **111**, 376 (1964).
54. Ives, D.J.G., Janz, G.J.: Reference electrodes. New York: Academic Press 1961.
55. Bates,R.G.: „Determination of pH." New York: J. Wiley 1964.
56. Henderson, P.: Z. physik. Chem. **63**, 325 (1908).
57. Planck, M.: Ann. Phys. Chem. N.F. **40**, 561 (1890).
58. MacInnes, D.A., Yeh, Y.L.: J. Am. Chem. Soc. **43**, 2563 (1921).
59. Butler, J.N.: Thermodynamic studies. In: Ion-selective electrodes, ed. R.A. Durst. Washington: NBS, Spec. Publ. 314, 1969.
60. Covington, A.K.: Reference electrodes. New York: Academic Press 1961.
61. Milazzo, G.: Elektrochemie. Wien: Springer 1952.
62. Harned, H.S., Owen, B.B.: Physical chemistry of electrolytic solutions. New York: Reinhold 1950.
63. Baucke, F.G.K.: J. Electroanal. Chem. **33**, 135 (1971).
64. Buzza, E.E., Leonard, J.E., Watanabe, H., Carlsen, E.N.: J. Assoc. Advancement of Medical Instr. **4**, 136 (1970).
65. Alner, D.J., Greczek, J.J., Smeeth, A.G.J.: J. Chem. Soc. **1205** (1967).
66. Manufacturer recommendation
67. Kostyuk, P.G., Sorokina, Z.A., Kholodova, Y.D.: In: Glass microelectrodes [253].
68. Bates, R.G., Alfenaar, M.: Activity standards for ion-selective electrodes. In: Ion-selective electrodes, ed. R.A. Durst. Washington: NBS, Spec. Publ. 314, 1969.
69. Ashby, J.H., Crook, E.M., Datta, S.P.: J. Biochem. **56**, 190 (1954).
70. Petersen, O.: Chemie-Ing.-Techn. **40**, 76 (1968).
71. Harned, H.S., Ehlers, R.W.: J. Chem. Soc. **55**, 2179 (1933).
72. Bates, R.G., Bower, V.E.: J. Res. NBS **53**, 283 (1954).
73. Manov, G.G. DeLollis, N.J., Acree, S.F.: J. Res. NBS **45**, 115 (1945).
74. Fricke, H.K.: In: Beiträge zur angewandten Glasforschung. Hrsg. E. Schott. Stuttgart: Wissenschaftl. Verlags GmbH. 1960.
75. Jenaer Glaswerke Schott & Gen., Mainz (Patent).
76. Belyustin, A.A., Lev, A.A.: Chem. in Natural Science, Leningrad State University **32** (1965).
77. Corning: Manual Na-Electrode.
78. Geyer, R., Erxleben, W., Syring, W.: Z. analyt. Chem. **204**, 325 (1964).
79. Rechnitz, G.A., Zamochnick, S.B.: Talanta **11**, 979 (1964).
80. Smith, M.J., Manahan, S.E.: Anal. Chem. **45**, 636 (1973).
81. Gomori, G.: Proc. Soc. exp.Biol. (N.Y.) **62**, 33 (1946).
82. Friedman, S.M., Bowers, F.K.: Analyt. Biochem. **5**, 471 (1963).
83. Leonard, J.E., Sibbald, P.G.: paper Pittburg conf. 1964.
84. Harzdorf, C.: Z. Analyt. Chem. **270**, 23 (1974).
85. Kolthoff, J.M., Sanders, H.L.: J. Am. chem. Soc. **59**, 416 (1937).

86. Radiometer: Bulletin 984 – 607 T.
87. Adametzová, H., Vadura, R.: Electroanal. Chem. and Interfac. Electrochemistry **55**, 53 (1974).
88. von Osch, G.W.S., Griepink, B.: Z. Analyt. Chem. **273**, 271 (1975).
89. Jaenicke, W.: Z. Elektrochem. **55**, 648 (1951).
90. Orion: Newsletter 9a. 10 (1970).
91. Mohan, M.S., Rechnitz, G.A.: Anal. Chem. **45**, 1323 (1973).
92. Lechner, J.F., Sekerka, I.: Electroanalyt. Chem. and Interfac. Electrochemistry **57**, 317 (1974).
93. Srinivasan, K., Rechnitz, G.A.: Anal. Chem. **40**, 509 (1968).
94. Buffle, J., Parthasarathy, N., Haerdi, W.: Anal. Chim. Acta **68**, 253 (1974).
95. Veselý, J., Stulik, K.: Anal. Chim. Acta **73**, 157 (1974).
96. Mertens, J., van den Winkel, P., Massart, D.L.: Anal. Chem. **48**, 272 (1976).
97. Durst, R.A., Taylor, J.K.: Anal. Chem. **39**, 1483 (1967).
98. Oehme, F.: NT-Automation u. Industrielle Elektronik **1**, 1 (1974).
99. Berndt, A.F., Stearns, R.I.: Anal. Chim. Acta **74**, 446 (1975).
100. Orion: Manual fluoride electrode (1971).
101. Haber, F.: Ann. Physik (4) **26**, 927 (1908).
102. Marshall, C.E.: J. Phys. Chem. **43**, 1155 (1939).
103. Tendeloo, H.J.C., Krips, A.: Rec. Trav. Chim. **76**, 703 (1957).
104. Pungor, E., Tóth, K.: Acta chim. Acad. Sci. hung. **41**, 239 (1964).
105. Pungor, E., Havas, J., Tóth, K.: Z. Chem. **5**, 9 (1965).
106. Pungor, E., Havas, J., Tóth, K.: Inst. Control System **38**, 105 (1965).
107. Pungor, E., Havas, J., Tóth, K.: Mikrochim. Acta 689 (1966).
108. Pungor, E.: Analyt. Chem. **39**, 28 A (1967).
109. Pungor, E., Schmidt, E., Tóth, K.: Proc. IMEKO Symposium on Electrochemical Sensors, Vesprem, Ungarn S. 121 (1968).
110. Mascini, M., Liberti, H.: Anal. Chim. Acta **60**, 405 (1972).
111. Geyer, R., Syring, W.: Z. Chem. **6**, 92 (1966).
112. Macdonald, G.M.A., Tóth, K.: Anal. Chim. Acta **41**, 99 (1968).
113. Shatkay, A.: Analyt. Chem. **39**, 1056 (1967).
114. Tendeloo, H.J.C., Voort, F.H. van der: Rec. Trav. Chim. **79**, 639 (1960).
115. Hirsch-Ayalon, P.: Polymer Sci. **23**, 697 (1957).
116. Fischer, R.B., Babcock, R.F.: Anal. Chem. **30**, 1732 (1958).
117. Tendeloo, H.J.C., Krips, A.: Rec. Trav. Chim. **78**, 177 (1959).
118. Morazzani-Pelletier, S., Baffier, M.: Chim. Phys. **62**, 429 (1965).
119. Wyllie, M.R.J., Patnode, W.: J. Phys. Chem. **54**, 204 (1950).
120. Helfferich, F.: Ionenaustauscher. Weinheim: Verlag Chemie 1959.
121. Malissa, H., Grasserbauer, M., Pungor, E., Tóth, K., Pápay, M.K., Polos, L.: Anal. Chim. Acta **80**, 223 (1975).
122. Rechnitz, G.A., Lin, Z.F., Zamochnick, S.B.: Analyt. Lett. **1**, 29 (1967).
123. Malissa, H., Jellinek, G.:Z. analyt. Chem. **245**, 70 (1969).
124. Sollner, K., Abrams, I., Carr, C.W.: J. gen. Physiol. **24**, 467 (1941).
125. Sollner, K.: Ann. N.Y. Acad. Sci. **148**, 154 (1968).
126. Ross, J.W.: Science **156**, 1378 (1967).
127. Ross, J.W.: Solid-state and liquid membrane ion-selective electrodes. In: ion-selective electrodes. ed. R.A. Durst. Washington: NBS, Spec. Publ. 314, 1969.
128. Ross, J.W., Frant, M.S.: paper, Pittsburg Conf. 1966.
129. Carlson, R.M., Paul, J.L.: Anal. Chem. **40**, 1992 (1968).
130. Moore, C., Pressman, B.C.: Biochem. biophys. Res. Commun. **15**, 562 (1964).
131. Simon, W., Pioda, L.A.R.: Chimia **23**, 72 (1969).
132. Simon, W.: Angew. Chem. **82**, 433 (1970).
133. Petránek, J., Ryba, O.: Anal. Chim. Acta **72**, 375 (1974).
134. Simon, W.: Vortrag, Analytika, München 1972.
135. Kedem, O., Perry, M., Bloch, R.: paper IUPAC-Symposium on selective ion-sensitive electrodes, Cardiff 1973.

136. Ammann, D., Bissig, R., Cimerman, Z., Fiedler, U., Güggi, M., Morf, W.E., Oehme, M., Osswald, H., Pretsch, E., Simon, W.: „Ion and Enzyme Electrodes in Biology and Medicine". Baltimore (U.S.A.) University Park Press 1976.
137. Moody, G.J., Thomas, J.D.R.: Selective Ion-Sensitive Electrodes, Watford: Merrow Publishing Co. Ltd., England 1971.
138. Moody, G.J., Oke, R.B., Thomas, J.D.R.: Analyst. **95**, 910 (1970).
139. Mascini, M., Pallozzi, F.: Anal. Chim. Acta **73**, 375 (1974).
140. Morf, W.E., Lindner, E., Simon, W.: Anal. Chem. **47**, 1596 (1975).
141. Llenado, R.A.: Anal. Chem. **47**, 2243 (1975).
142. Hammond, S.M., Lambert, P.A.: Electroanalyt. Chem. and Interfac. Electrochemistry **53**, 155 (1974).
143. Baczuk, R.J., DuBois, R.J.: Anal. Chem. **40**, 685 (1968).
144. Hirsch, R.F., Portock, J.D.: Anal. Lett. **2**, 295 (1969).
145. Schultz, F.A., Mathis, D.E.: Anal. Chem. **46**, 2253 (1974).
146. Hirsch, R.F., Olderman, G.M.: Anal. Chem. **48**, 771 (1976).
147. Herman, H.B., Rechnitz, G.A.: Anal. Chim. Acta **76**, 155 (1975).
148. Szczepaniak, W., Ren, K.: Anal. Chim. Acta **82**, 37 (1976).
149. Baumann, E.W.: Anal. Chem. **48**, 548 (1976).
150. Matsui, M., Freiser, H.: Anal. Lett. **3**, 161 (1970).
151. Nagelberg, I., Bradock, C.I., Barbero, G.J.: Science **166**, 1403 (1969).
152. Scibona, G., Mantella, L., Danesi, P.R.: Anal. Chem. **42**, 844 (1970).
153. Haynes, W.M., Wagenknecht, J.H.: Anal. Letters **4**, 491 (1971).
154. Catrall, R.W., Pin, C.P.: Anal. Chem. **48**, 552 (1976).
155. Tanaka, T., Hiiro, K., Kawahara, A.: Anal. Lett. 7, 173 (1974).
156. Gavach, C., Bertrand, C.: Anal. Chim. Acta **55**, 385 (1971).
157. Levins, R.J.: Anal. Chem. **43**, 1045 (1971).
158. Entwistle, J.R., Hayes, T.J.: paper IUPAC-Symposium on selective ion-sensitive electrodes, Cardiff 1973.
159. Ruzicka, J., Lamm, C.G., Tjell, J.: Anal. Chim.Acta **62**, 15 (1972).
160. Ansaldi, A., Epstein, S.I.: Anal. Chem. **45**, 595 (1973).
161. Simon, W., Morf, W.E.: paper-International Workshop on Ion Selective Electrodes and on Enzyme Electrodes in Biology and in Medicine, Schloß Reisenberg, Germany 1974.
162. James, H., Carmack, G., Freiser, H.: Anal. Chem. **44**, 856 (1972).
163. Cattrall, R.W., Pui Chin-Poh: Anal. Chem. **47**, 93 (1975).
164. Fujinaga, T., Okazaki, S., Freiser, H.: Anal. Chem. **46**, 1842 (1975).
165. Stow, R.W., Baer, R.F., Randall, B.F.: Archs. Phys. Med. Rehabil. **38**, 646 (1957).
166. Severinghaus, W., Bradley, A.F.: J. Appl. Physiol. **13**, 515 (1958).
167. Ruzicka, J., Hansen. E.H.: Anal. Chim. Acta **69**, 129 (1974).
168. Bailey, P.L., Riley, M.: Analyst **100**, 1188 (1975).
169. Bailey, P.L.: Analyses with Ion-Selective Electrodes, London, Heyden 1976
170. Ross, J.W., Riseman, J.H., Krueger, J.A.: Plenary Lecture-IUPAC-Symposium on Selective Ion-Sensitive Electrodes, Cardiff 1973.
171. Anfält, T., Graneli, A., Jagner, D.: Anal. Chim. Acta **76**, 253 (1975).
172. Himpler, H.: personal communication
173. Rechnitz, G.A.: CENEAR 53 (4), 29 (1975).
174. Moody, G.J., Thomas, J.D.R.: Analyst **100**, 609 (1975).
175. Katz, S.A., Rechnitz, G.A.: Z. analyt. Chem. **196**, 248 (1963).
176. Clark, L.C., Lyons, L.: Ann. New York Acad. of Science **102**, 29 (1962).
177. Hicks, G.P., Updike, S.K.: Anal. Chem. **38**, 726 (1966).
178. Guilbault, G.G., Montalvo, J.G.: J. Amer. chem. Soc. **91**, 2164 (1969).
179. Guilbault, G.G., Montalvo, J.G.: J. Amer. chem. Soc. **92**, 2533 (1970).
180. Guilbault, G.G., Hrabankova, E.: Anal. Chem. **42**, 1779 (1970).
181. Guilbault, G.G.: Anal. Chim. Acta **56**, 285 (1971).
182. Tran-Minh, C., Brown, G.: Anal. Chem. **47**, 1359 (1975).

183. Papastathopoulos, D.S., Rechnitz, G.A.: Anal. Chim. Acta **79**, 17 (1975).
184. Guilbault, G.G., Stokbro, W.: Anal. Chim. Acta **76**, 237 (1975).
185. Jerfy, A., Roy, A.B.: Analyt. Biochem. **49**, 610 (1972).
186. Papastathopoulos, D.S.: personal communication (1975).
187. Papastathopoulos, D.S., Rechnitz, G.A.: Anal. Chem. **47**, 1792 (1975).
188. Richmond, W.: Clin. Chem. **19**, 1350 (1973).
189. Janata, J.: J. Am. Chem. Soc. **97** (10), 2914 (1975).
190. Hansen, E.H., Ruzicka, J.: Anal. Chim. Acta **72**, 353 (1974).
191. Morris, D.L., Campell, J., Hornby, W.E.: Biochem. J. **147**, 593 (1975).
192. Baum, G., Ward, F.B.: Analyt. Biochem. **42**, 487 (1971).
193. Storp, L.H., Guilbault, G.G.: Anal. Chim. Acta **62**, 425 (1972).
194. Crochet, K.L., Montalvo, J.G.: Anal. Chim. Acta **69**, 129 (1974).
195. Gibson, K., Guilbault, G.G.: Anal. Chim. Acta **76**, 245 (1975).
196. Alexander, P.W., Rechnitz, G.A.: Anal. Chem. **46**, 250, 860 (1974).
197. Llenado, R.A., Rechnitz, G.A.: Anal. Chem. **43**, 1457 (1971).
198. Mascini, M., Liberti, A.: Anal. Chim. Acta **68**, 177 (1974).
199. Larsen, N.R., Hansen, E.H., Guilbault, G.G.: Anal. Chim. Acta **79**, 9 (1975).
200. Guilbault, G.G., Smith, R., Montalvo, J.: Anal. Chem. **41**, 600 (1969).
201. Llenado, R.A., Rechnitz, G.A.: Anal. Chem. **44**, 1366 (1972).
202. Guilbault, G.G., Gutknecht, W.F., Kuan, S.S., Cochran, R.: Analyt. Biochem. **46**, 200 (1972).
203. Erlanger, B.F., Sack, R.A.: Analyt. Biochem. **33**, 318 (1970).
204. Llenado, R.A., Rechnitz, G.A.: Anal. Chem. **45**, 2165 (1973).
205. Guilbault, G.G., Tarp, M.: Anal. Chim. Acta 73, 355 (1974).
206. Guilbault, G.G., Shu, F.R.: Anal. Chem. **44**, 2161 (1972).
207. Thompson, H., Rechnitz, G.A.: Anal. Chem. **46**, 246 (1974).
208. Hussein, W.R., Guilbault, G.G.: Anal. Chim. Acta **72**, 381 (1974).
209. Kiang, C.H., Kuan, S.S., Guilbault, G.G.: Anal. Chim. Acta **80**, 209 (1975).
210. Cullen, L.F., Rusling, J.F., Schleifer, A., Papariello, G.J.: Anal. Chem. **46**, 1955 (1974).
211. Guilbault, G.G., Nagy, G.: Anal. Lett. **6**, 301 (1973).
212. Crombie, D.J., Moody, G.J., Thomas, J.D.R.: in preparation.
213. Cammann, K.: Elektronik 1975 (4), 79 (1975).
214. Bitterlich, W.: Einführung in die Elektronik. Berlin-Heidelberg-New York: Springer 1967 und „electrometer measurements", Keithley Instruments 1972.
215. Papastathopoulos, D.S., Rechnitz, G.A.: Anal. Chem. **48**, 862 (1976).
216. Handbook of analytical chemistry, ed. L. Meites. New York: McGraw-Hill Book Co., Inc., – effective ionic radii: Kielland. J. Amer. chem. Soc. **59**, 1675 (1937).
217. Orion: Manual: Cu-Electrode.
218. Kraft, G., Fischer, H.: „Indikation von Titrationen". Berlin: W. de Gruyter & Co. 1972.
219. Carr, P.W.: Analyt. Chem. **43**, 425 (1971).
220. Carr, P.W.: Analyt. Chem. **44**, 452 (1972).
221. Anfält, T., Jagner, D.: Anal. Chem. **45**, 2412 (1973).
222. Leithe, W.: Chem.-Ing.-Techn. **36**, 112 (1964).
223. Oehme, F., Dolzalova, L.: Z. analyt. Chem. **264**, 168, (1973).
224. Oehme, F., Dolzalova, L.: Z. analyt. Chem. **251**, 1 (1970).
225. Durst, R.A.: Ion-selective electrodes. Washington: NBS, Spec. Publ. 314, 1969.
226. Khuri, R.N.: Ion-selective electrodes in biomedical research. In: Ion-selective electrodes, ed. R.A. Durst. Washington: NBS, Spec. Publ. 314 1969.
227. Orion: Newsletter 5 and 6 1970.
228. Orion: Newsletter 7 and 8 1970.
229. Orion: Newsletter 2 1970.
230. Gran, G.: The Analyst 77 (11), 661 (1952).
231. Orion Kat. Nr. 90-00-90.
232. Beckman-S75157-875-5DC
233. Orion: Newspaper 1 a. 2 (1971).
234. Schwarzenbach, G., Flaschka, H.: Die komplexometrische Titration. Stuttgart: F. Enke 1965.

235. Orion: Newspaper 3 a. 4 (1971).
236. Eriksson, T.: Anal. Chim. Acta **58**, 437 (1972).
237. Burden, S.L., Euler, D.E.: Anal. Chem. **47**, 793 (1975).
238. Isbell, Jr., A.F., Pecsok, R.L., Davies, R.H., Purnell, J.H.: Anal. Chem. **45**, 2363 (1973).
239. Anderson, K.P., Butler, E.A., Wolley, E.M.: J. phys. Chem. **75**, 93 (1971).
240. Krijgsman, W., Mansveld, J.F., Griepink, B.: Z. analyt. Chem. **249**, 368 (1970).
241. Abresch, K., Claasen, I.: Die coulometrische Analyse. Weinheim: Verlag Chemie 1961.
242. McLean, F.C., Hastings, A.B.: J. biol. Chem. **107**, 337 (1934).
243. McLean, F.C., Hastings, A.B.: Amer. J. Med. Sci. **189**, 601 (1935).
244. Moore, E.W.: In: Ion-selective electrodes, ed. R.A. Durst. Washington: NBS, Spec. Publ. 314, 1969.
245. Krück, F.: Kalium u. Elektrolythaushalt, Sonderdruck v. Gebr. Giulini GmbH, Ludwigshafen 1966.
246. Lustgarten, J.A., Wenk, R.E., Byrd, C., Hall, B.: Clin. Chem. **20**/9, 1217 (1974).
247. Kruse-Jarres, J.D., Noeldge, G.: Ärztl. Lab. **21**, 259 (1975).
248. Fuchs, C., Dorn, D., McIntosh, C.: Z. analyt. Chem. **279**, 150 (1976).
249. Friedman, S.M.: In: The glass electrode, eds. G. Eisenman, R. Bates, G. Mattock and S.M. Friedman. New York: Interscience 1966.
250. Perkins, H.A., Snyder, M., Tacher, C., Roles, M.R.: Transfusion **11**, 204 (1971).
251. Caldwell, P.C.: J. Physiol. (Lond.) **142**, 22 (1958).
252. Hinke, J.A.M.: J. Physiol. (Lond.) **156**, 314 (1961).
253. Glass microelectrodes, eds.: M. Lavallée, O.F., Schanne, N.C. Hébert. New York: J. Wiley 1969.
254. Lev, A.A.: Nature **201**, 1132 (1964).
255. „Ion and Enzyme Electrodes in Biology and Medicine". Baltimore (U.S.A.). University Park Press 1976.
256. Kessler, M.: personal communication.
257. Silver, I.A.: paper-International Workshop on Ion Selective Electrodes and on Enzyme Electrodes in Biology and Medicine, Schloß Reisenberg, Germany 1974 [255].
258. Walker, J.L.: Anal. Chem. **43**, 89 A (1971).
259. Neher, E., Lux, H.D.: Messung extrazellulärer K-Anreicherung während eines „voltage camp pulses". Referate d. Deutschen Physiolog. Ges., 39. Tag., Erlangen. Berlin-Heidelberg- New York: Springer 1972.
260. Šucha, L., Suchánek, M.: Anal. Lett. **3**, 613 (1970).
261. U.S. Geol. Survey, Water Supply Paper 1827-C.
262. Jaselskis, B., Bandemer, M.K.: Anal. Chem. **41**, 855 (1969).
263. Baumann, E.W.: Anal. Chem. **42**, 110 (1970).
264. Orion: Analytical methods guide, 5. edition 1972.
265. Bauwart, W.L. Tabatabai, M.A., Bremner, J.M.: Comm. in Soil and Plant Anal. **3**, 449 (1972).
266. Heyn, A.H., Bergin, M.J.: J. Am. Chem. Soc. **75**, 5120 (1953).
267. Shiurba, R.A., Jolly, W.L.: J. Am. Chem. Soc. **90**, 5289 (1968).
268. Bottazzini, N., Crespi, V.: La Chimica e L'Industria, **52**, 866 (1970).
269. Mascini, M., Liberti, A.: Anal. Chim. Acta **60**, 405 (1972).
270. Ross, J.W., Frant, M.S.: Anal. Chem. **41**, 1900 (1969).
271. Rechnitz, G.A., Kenny, N.C.: Anal. Lett. **3**, 259 (1970).
272. Frant, M.S.: Plating, July 1971, 686 (1971).
273. Carlson, R.M., Paul, J.L.: Anal. Chem. **40**, 1292 (1968).
274. Carlson, R.M., Paul, J.L.: Soil Sci. **108**, 266 (1969).
275. Potman, W., Dahmen, E.: Mikrochim. Acta 1972, 303 (1972).
276. Harris, R.C., Williams, H.H.: J. Appl. Meteorol. **8**, 299 (1969).
277. Ficklin, W.H., Gotschall, W.C.: Anal. Lett. **6**, 217 (1973)
278. Turner, D.L.: J. Food Sci. **37**, 791 (1972).
279. Cammann, K.: Naturwissenschaften **57**, 298 (1970).
280. Woolson, E.A., Axley, J.H., Kearney, P.C.: Soil Sci. **109**, 279 (1970).
281. Thompson, M.E., Ross, J.W.: Science **154**, 1643 (1966).
282. Kester, D.R., Rytkowicz, R.M.: Limn & Oceanog. **14**, 686 (1969).

283. Arnold, D.E., Stausell, M.J., Malvin, H.H.: Am. J. of Clin. Path. **49**, 627 (1968).
284. Moore, E.W., Blum, A.L.: J. Clin. Invest. **47**, 70a (1968).
285. Bernstein, D.S., Aliapoulious, M.A., Hattner, R.S., Wachman, A., Rose, B.: Endocrinology **55**, 589 (1969).
286. Moore, E.W.: J. Clin. Invest. **49**, 318 (1970).
287. Harris, E.K., DeMets, D.L.: Clin. Chem. **17**, 983 (1971).
288. Hattner, R.S., Johnson, J.W., Bernstein, D.S., Wachman, A., Brackman, J.: Clin. Chim. Acta **28**, 67 (1970).
289. Sachs, C., Bourdeau, A.M., Broyer, M., Balsan, S.: Ann. Biol. Clin. **28**, 137 (1970).
290. Lindgarde, F.: Clin. Chim. Acta **40**, 477 (1972).
291. Oreskes, I., Douglas, K.S.: Am. Rheumatism Ass. Meeting 502 (1968).
292. Oreskes, I., Hirsch, C., Douglas, K.S., Kupfer, S.: Clin. Chim. Acta **21**, 303 (1968).
293. Pittinger, C.B.: Anes. & Anal. **49**, 540 (1970).
294. Radde, I.C., Höffgen, B., Parkinson, D.K., Sheepers, J., Luckham, A.: Clin. Chem. **17**, 1002 (1971).
295. Li, T.K., Piechocki, J.T.: Clin. Chem. **17**, 411 (1971).
296. Raman, A.: Biochem. Med. **3**, 369 (1970).
297. Robertson, W.G., Peacock, M.: Clin. Chim. Acta **20**, 315 (1968).
298. Seamonds, B., Towfighi, J., Arvan, D.A.: Clin. Chem. **18**, 155 (1972).
299. Studer, O., Knob, M., Binswanger, U.: Schweiz. Med. Wochenschr. **102**, 305 (1972).
300. Stoltz, W.L., Nolph. K.D., Maher, J.F.: J. Lab. Clin. Med. **78**, 389 (1971).
301. Höffgen, B., Parkinson, D.K., Storms, P., Radde, I.C.: Clin. Orthop. **78**, 30 (1971).
302. Sachs, C.E., Bourdeau, A.M.: Clin. Orthop. **78**, 24 (1971).
303. Fuchs, C., Paschen, K.: Deut. Med. Wochenschr. **7**, 23 (1972).
304. Blum, J.N., Ramberg, C.F., Johnson, K.G., Kronfeld, D.S.: Am. J. Vet. Res. **33**, 51 (1972).
305. Sachs, C.E., Bourdeau, A.M.: Rev. Europ. Etudes Clin. et Biol. **17**, 218 (1972).
306. Lin, K.L., Beng, C.G., Lam, K.S.: Proc. of the 5th Malaysia-Singapore Congress of Medicine **5**, 191 (1970).
307. Brown, D.M., Boen, J., Bernstein, A.: Pediatrics **49**, 841 (1972).
308. Pittinger, C., Chang, P.M., Faulkner, W.: Southern Med. J. **64**, 1211 (1971).
309. Killen, D.A., Grogau, E.L., Gower, R.E., Collins, H.A.: Surgery **70**, 736 (1971).
310. Baman, A.: Clin. Biochem. **4**, 141 (1971).
311. Lindgärte, F., Zettervall, O.: Israel J. Med. Sci. **7**, 510 (1971).
312. Hansen, S.O., Theodorsen, L.: Clin. Chim. Acta **31**, 119 (1971).
313. Schwartz, H.D., McConville, B.C., Christophersen, E.F.: Clin. Chim. Acta **31**, 97 (1971).
314. Hinkle, J.E., Cooperman, L.H.: Brit. J. Anaesth. **43**, 1108 (1971).
315. Bourdeau, A.M., Sachs, C., Presle, V., Dromini, M.: Pharm. Biol. **67**, 527 (1971).
316. Sachs, C.: Presse Medicale **78**, 1547 (1970).
317. Sachs, C., Bourdeau, A.M.: J. Physiol. (Paris) suppl. **62**, 313 (1970).
318. Braddock, L.I., Nagelberg, I., Margallo, E., Barbero, G.J.: Cystic Fibrosis Club Abstract, April 1969, 17 (1969).
319. Grøn, P., Spinelli, M.A., Hag, D.I.: IADR Abstracts **58** (1958).
320. Moore, E.W., Makhlouf, G.M.: Gastroenterology **55**, 465 (1968).
321. Urist, M.R., Speer, D.P., Ibsen, K.J., Strates, B.S.: Calc. Tiss. Res. **2**, 253 (1968).
322. Woodward, C., Davidson, E.A.: Proc. Nat. Acad. Sci. U.S.A. **60**, 201 (1968).
323. Chance, B., Yoshioka, T.: Biochem. **5**, 3224 (1966).
324. Perris, A.D., Whithield, J.F., Tolg, P.K.: Nature Lond. **219**, 527 (1968).
325. Sachs, C., Bourdeau, A.M., Presle, V.: Rev. Europ. Etudes Clin. et Biol. **16**, 831 (1971).
326. Orion: Newsletter Vol. III, 9 und 10 (1971).
327. Bourdeau, A.M., Sachs, C.: Le Pharmacien Biologiste **6**, 527 (1970).
328. Jacobs, J.S., Hattner, R.S., Bernstein, D.S.: Clin. Chim. Acta **31**, 467 (1971).
329. Muldoon. P.J., Liska, B.J.: Dairy Sci. **52**, 460 (1969).
330. Krämer, R., Lagoni, H.: Naturwissenschaften **56**, 36 (1969).
331. Tackett, S.L.: Anal. Chem. **41** (1969).
332. Hadjiioannon, T.P., Papastathopoulos, D.S.: Talanta **17**, 399 (1970).

333. Christiansen, T.F., Busch, J.E., Krogli, S.C.: Anal. Chem. **48**, 1051 (1976).
334. Henscheid, T., Schoenrock, K., Berger, P.: J. of the A.S.S.B.T. **16**, 482 (1971).
335. Florence, T.M.: J. Electroanal. Chem. **31**, 77 (1971).
336. Haynes, S.J., Clark, A.H.: Economic Geology **67**, 378 (1972).
337. Hipp, B.W., Langdale, G.W.: Soil Sci. and Plant Anal. **2**, 237 (1971).
338. Carlson, R.M., Keeny, D.R.: Instrumental Methods for Analysis of Soil and Plant Tissue 39 (1971).
339. Randell, A.W., Linklater, P.M.: Australien J. Dairy Tech. **27**, 51 (1972).
340. Lee, T.G.: Anal. Chem. **41**, 391 (1969).
341. Mildoon, P.J., Liska, B.J.: J. Dairy Sci. **54**, 117 (1971).
342. LaCroix, R.L., Keeny, D.R., Walsh, L.M.: Soil Sci. and Plant Anal. **1**, 1 (1970).
343. Warner, T.B.: Marine Technology Soc., 6th Annual Reprints **2**, 1495 (1970).
344. Ogata, N.: Jap. Analyst **21**, 780 (1972).
345. Jagner, D., Arén, K.E.: Anal. Chim. Acta **52**, 491 (1970).
346. Kriigsman, W., Mansveld, J.F., Griepink, B.: Clin. Chim. Acta **29**, 575 (1970).
347. Kopito, L., Schwachman, H.: Pediatrics **43**, 794 (1969).
348. Armstrong, D., Kramer, J.C.: paper XII Annual Meeting of the Cystic Fibrosis Club, 28.4.1971 Atlantic City (1971).
349. Szabo, L., Kenny, M.A., Lee, W.: Clin. Chem. **19**, 727 (1973).
350. Frant, M.S., Ross, J.W., Riseman, J.H.: Anal. Chem. **44**, 2227 (1972).
351. Riseman, J.H.: Am. Lab. **4**, 63 (1972).
352. Oehme, F.: J. für Oberflächentechnik **6**, 5 (1974).
353. Gillinghain, J.T., Shirer, M.M., Page, N.R.: Agron. J. **61**, 717 (1969).
354. Vickroy, G.G., Gaunt, G.L.: Tobacco **174**, 50 (1972).
355. Frant, M.S.: Plating **54**, 702 (1967).
356. Oehme, F.: CZ-Chemie-Technik, **3**, 27 (1974).
357. Elfers, L.A., Decker, C.E.: Anal. Chem. **40**, 1658 (1968).
358. Yunghans, R.S., McMullen, T.B.: Fluoride **3**, 143 (1970).
359. Buck, M., Reusmann, G.: Fluoride **4**, 5 (1971).
360. Thompson, R.J., McMullen, R.J., Morgan, G.B.: J. of Air Pollution Control Ass. **21**, 484 (1971).
361. Farzaneh, A., Troll, G., Cammann, K.: Chemical Geology **20**, 295 (1977)
362. Warner, T.B.: Science **165**, 178 (1969).
363. Warner, T.B.: Deep Sea Research **18**, 1255 (1971).
364. Windom, H.L.: Limn. and Oceanog. **16**, 806 (1971).
365. Warner, T.B.: Geophys. Res. 77, 2728 (1972).
366. Ballantyne, D.J.: Atmos. Envir. **6**, 267 (1972).
367. Torma, L., Ginther, B.E.: J. Assoc. Offic. Agr. Chemists **51**, 1181 (1968).
368. Tusl, J.: J. Assoc. Offic. Agr. Chemists **53**, 267 (1970).
369. Croomes, E.F., McNutt, R.C.: Analyst **93**, 729 (1968).
370. Fry, B.W., Taves, D.R.: J. Lab. Clin. Med. **75**, 1020 (1970).
371. Tusl, J.: Clin. Chim. Acta **27**, 216 (1970).
372. Neefus, J.D., Cholack, J., Saltzmann, B.E.: Am. Ind. Hyg. Assoc. J. **31**, 96 (1970).
373. Taves, D.R.: Nature **217**, 1050 (1968).
374. Taves, D.R.: Talanta **15**, 1015 (1968).
375. Venkateswarly, P., Singer, L., Armstrong, W.D.: Anal. Biochem. **42**, 350 (1971).
376. Brudevold, F., McCann, H.G., Grøn, P.: Arch. Oral. Biol. **13**, 877 (1968).
377. Hotz, P.: Helv. Odontre Acta **16**, 32 (1972).
378. Shane, N., Miele D.: J. Pharm. Sci. **57**, 1260 (1968).
379. Bushee, E.J., Grissom, D.K., Smith, D.R.: J. Dent. Child. **38**, 279 (1971).
380. Grøn, P., Yao, K., Spinelli, M.: J. Dent. Res. Suppl. to No 5, 48, 709 (1969).
381. Weiss, D.: Z. Analyt. Chem. **262**, 28 (1972).
382. Hozumi, K., Akimoto, N.: Anal. Chem. **42**, 1312 (1970).
383. Hoover, W.L., Melton, J.R., Howard, P.A.: J. Assoc. Offic. Agr. Chemists **54**, 760 (1971).
384. Paletta, B., Panzenbeck, K.: Clin. Chim. Acta **26**, 11 (1969).
385. Paletta, B.: Mikrochim. Acta **6**, 1210 (1969).

386. Arino, H., Kramer, H.H.: Nucl. Appl. **4**, 356 (1968).
387. Westerlund-Helmerson, U.: Anal. Chem. **43**, 1180 (1971).
388. Woodson, J.H., Liebhafsky, H.A.: Nature **224**, 690 (1969).
389. Lingane, J.J.: Anal. Chem. **39**, 881 (1967).
390. Lingane, J.J.: Anal. Chem. **40**, 935 (1968).
391. Baumann, E.W.: Anal. Chem. **40**, 1731 (1968).
392. Sander, C., Ts'o, P.O.P.: J. Mol. Biol. **55**, 1 (1971).
393. Thompson, M.E.: Science **153**, 867 (1966).
394. Kester, D.R., Pytkowicz, R.M.: Limn. and Oceanog. **13**, 670 (1968).
395. Gruen, L.C., Harrap, B.S.: Anal. Biochem. **42**, 377 (1971).
396. Rechnitz, G.A., Mohan, M.S.: Science **168**, 1460 (1970).
397. Frant, M.S., Ross, J.W.: Science **167**, 987 (1970).
398. Eckfeldt, E.L., Procter, Jr.N.E.: Anal. Chem. **47**, 2307 (1975).
399. Cammann, K.: Fortschr. d. chem. Forschung **8**, (2), 222 (1967).
400. Green, M., Behrendt, H., Libien, G.: Clin. Chem. **18**, 427 (1972).
401. Lenz, B.L., Mold, J.R.: TAPPI, 54, 2051 (1971).
402. Annino, J.S.: Clin. Chem. **13**, 227 (1967).
403. Mason, W.D., Needham, T.E., Price, J.C.: J. Pharm. Sci. **60**, 1756 (1971).
404. Langmuir, D., Jacobson, R.L.: Envir. Sci. and Tech. **4**, 834 (1970).
405. Bremner, J.M., Bundy, L.G., Agarwal, A.S.: Anal. Lett. **13**, 837 (1968).
406. Myers, R.J.K., Paul, E.A.: Can. J. Soil Sci. **48**, 369 (1968).
407. Øien, A., Selmer-Olsen, A.R.: Analyst **94**, 888 (1969).
408. Mahendrappa, M.K.: Soil Sci. **108**, 132 (1969).
409. Mack, A.R., Sanderson, R.B.: Can. J. Soil Sci. **51**, 95 (1971).
410. Kelly, J.F., Bliss, D.W.: Hort Sci. 6 Section 2,1 (1971).
411. Barker, A.V., Peck, N.H., MacDonald, G.E.: Agron. J. **63**, 126 (1968).
412. Paul, J.L., Carlson, R.M.: J. Agr. Food Chem. **16**, 766 (1968).
413. Baker, A.S., Smith, R.: J. Agr. Food Chem. **17**, 1284 (1969).
414. Cantliffe, D.J., MacDonald, G.E., Peck, N.H.: N.Y. Food and Life Sci. Bull. **3**, Plant Sciences No 1 (1970).
415. Shaw, E.C., Wiley, P.: Calif. Agr. **5**, 11 (1969).
416. Rechnitz, G.A., Kenny, N.C.: Anal. Lett. **3**, 509 (1970).
417. Selig, W.: Mikrochem. J. **15**, 452 (1970).
418. Trachtenberg, A.F., Suffet, I.H.: paper-65th Annual Meeting of the Air Pollution Control Assoc. 18–22.6.1972.
419. Selig, W., Crossmann, G.L.: Informal Report UCID 15623, Lawrence Radiation Lab., Livermore, Calif.
420. Baczuk, R.J., Dubois, R.J.: Anal. Chem. **40**, 685 (1968).
421. Müller, D.C., West, P.W., Müller, R.H.: Anal. Chem. **41**, 2038 (1969).
422. Ross, J.W., Frant, M.S.: Anal. Chem. **41**, 967 (1969).
423. Goertzen, J.O., Oster, J.D.: Soil Sci. Soc. Am. Proc. **36**, 691 (1972).
424. Selig, W.: Mikrochim. Acta **1**, 168 (1970).
425. Heistand, R.N., Blake, C.T.: Mikrochim. Acta **3**, 212 (1972).
426. Brunow, G., Ilus, T., Miksche, G.E.: Acta Chemica Scand. **26**, 1117 (1972).
427. Harrap, B.S., Gruen, L.C.: Anal. Biochem. **42**, 398 (1971).
428. Movie, G.P.: Tobacco Sci. **107**, 34 (1971).
429. Oliver, R.T., Mannion, R.F.: ISA Anal. Instr. (preprint) 8, VIII–3 (1970).
430. Martin, C., Poudon, A.M.: Travaux de la Société de Pharmacie de Montepellier **31**, 371 (1971).
431. Fiedler, U., Hansen, E.H., Ruzicka, J.: Anal. Chim. Acta **74**, 423 (1975).
432. Cammann, K.: Messtechnik 3/71, 79 (1971).
433. Pioda, A.R., Simon, W., Bosshard, H.R., Curtius, H.Ch.: Cin. Chim. Acta **29**, 289 (1970).
434. Gilbert, T.R., Clay, A.M.: Anal. Chem. **45**, 1757 (1973).
435. Forney, L.J., McCoy, J.F.: Analyst **100**, 157 (1975).
436. van den Winkel, P., Mertens, J., Massart, D.L.: Anal. Chem. **46**, 1765 (1974).
437. Orion: Newsletter V, 1 (1973).

438. Galster, H.: Z. Analyt. Chem. **245**, 62 (1969).
439. Light, T.S. in: Ion-selective electrodes, ed. R.A. Durst. Washington: NBS Spec. Publ. 314, 1969.
440. Hirata, H., Higashiyama, K.: Talanta **19**, (1972).
441. Bergvelt, P.: IEEE Trans., BME–19, 342 (1972).
442. Matsuo, T., Wise, K.D.: IEEE Trans., BME–21, 485 (1974).
443. Moss, S.D., Janata, J., Johnson, C.C.: Anal. Chem. **47**, 2238 (1975).
444. Bockris, J.O.'M., Reddy, A.K.N.: Modern Electrochemistry. New York: Plenum Press 1970.
445. Durst, R.A.: paper IUPAC-Symposium on selective ion-sensitive electrodes, Cardiff 1973.
446. Bates, R.G.: paper IUPAC-Symposium on selective ion-sensitive electrodes, Cardiff 1973.
447. Robinson, R.A., Bates, R.G.: Anal. Chem. **45**, 1666 (1973).
448. Cammann, K.: Conference on Ion-Selective Electrodes Budapest 1977, P. 297, ed. Pungor, E., Budapest: Akadémiai Kiado 1978.
449. Tables of Standard Electrode Potentials, ed. Milazzo, G., Caroli, S. New York: J. Wiley & Sons 1978.
450. Trassatti, S.: Electroanal. Chem. and Interfacial Chem. **52**, 313 (1974).
451. Gomer, R.: Private Communication, Paper presented 1978 at the University of Munich.
452. Bloch, R., Shatkay, A., Saroff, H.A.: Biophys. J. **7**, 865 (1967).
453. Cammann, K.: Anal. Chem. **50**, 936 (1978).
454. Cammann, K.: Fresenius Z. anal. Chem. **287**, 1 (1977).
455. Pure Appl. Chem. 129 (1976).
456. Liteanu, C., Hopîrtean, E., Popescu, C.: Anal. Chem. **48**, 2013 (1976).
457. Liteanu, C., Hopîrtean, E., Popescu, C., Rîca, I., Stefaniga: Anal. Chem. **50**, 1202 (1978).
458. Madsen, S., Ladefoged, J.: Ugeskr. Laeg. 140 (14), 774 (1978).
459. Oehme, M.: Dissertation 5953, ETH Zürich 1977.
460. Cammann, K.: Chapter in: Analytiker Taschenbuch 1979, ed. Kienitz, Bock, Fresenius, Huber, Tölg. Berlin-Heidelberg-New York: Springer 1979.

Subject Index

Accuracy, activity determination 123
–, concentration determination 124
–, direct read off method 126
–, EMF measurement 112
–, gas sensors 98
–, industrial direct potentiometry 188
–, standard addition method 140, 145
–, titration method 127
Acetylcholine electrode 98, 100
Activity, calibration curve 122
–, coefficient 14, 162
––, determination 123
––, mean and individual 197
–, standards 199
Adenosine monophosphate sensor 100
Adsorption effects 73
Air gap electrode 94
Albumin sensor 100, 172
Aliquat 336 S 88
Alkali electrode 51, 77, 79
– error, pH electrodes 21
Aluminum determination 66, 163
––, industrial flow-thru 188
Amino acid electrodes 91
Ammonia determination 163
– electrode 93
Ammonium determination 163
––, in serum 98
– electrode 79
Amygdalin sensor 100
Analysis techniques 121
––, flow-thru measurements 186
Anion selective electrodes 79
Antibiotics 28, 79
Antibody sensor 102
Applications, ion-selective electrodes 162
Arginine sensor 100
Arsenic determination 163
Asparagine sensor 100
Asymmetric titration curve 130
Asymmetry control 47, 125
– potential 47
Automatic temperature control 126
Azide determination 163

Barium determination 163
– electrode 79, 84
Benzoate electrode 91
Beryllium determination 163
Bio-sensors 98
Blood analysis 172
Boron determination 163
Bromide determination 164
– electrode 63, 74, 76, 91
Bromine determination 164

Cable, insulation resistance 116
–, shielding 117
Cadmium determination 164
– electrode 58, 70, 76
Calcium determination 164
–– in biological samples 171
– electrode 74, 77, 84
Calibration 122
– curve 55
Calomel reference electrode 48
Cancer, diagnosis by electrode measurements 172
Capacitance neutralization 179
Carbon determination 165
Carbon dioxide electrode 93
Carbonate determination 165
– electrode 88
Carboxylate electrode 87
Catechol sensor 100
Cationic sensitive glass membrane electrodes 56
Cell without liquid junction 42, 191
Cerium determination 66
Cesium determination 165
– electrode 79, 88
Charging problems 116
Chemical potential 12, 16, 197
Chloride determination 157, 165
– electrode 63, 65, 74, 76, 79, 91
Chlorine determination 165
Cholesterol sensor 100
Chromate electrode 88
Chromium determination 165

Circuit current 109
Citrate determination 165
Clinical analyzers 173
Coated wire electrodes 89
Cobalt determination 165
Common mode rejection 120
Complexing agents determination 166
Complex bound ion determination 138
– formation constant 128
Concentration calibration curve 123
– cell 133, 186
– measurement 123, 126
Conditioning, electrode 77, 80, 90
– solution 73, 124, 140, 159, 180
Conductivity 105, 144
Construction of electrodes 49, 51, 59, 61, 74, 80, 82, 94, 99
Continuous measurements 179
Copper determination 166
– electrode 70, 76, 79
Creatinine sensor 101
Critical micell concentration measurement 88
Current-voltage curve 23
Cyanate determination 166
Cyanide determination 166, 188
– electrode 64, 74, 76

Dead-stop titration 132
Debye-Hückel theory 175, 198
Detection limit 66, 73, 78, 122, 141, 144, 158
Detergent electrode 88
Deviation from Nernstian behavior 121, 176
Diaphragm resistance 107
Differential input amplifier 187
Diffuse electric double layer 8
Diffusion potential 24, 37, 83
Direct potentiometry 121, 125, 187
Divalent cation electrode 84
Donnan potential 24
Drift of potential 16, 81, 90, 97, 107, 110, 123, 135, 174, 178

Electrical resistance of electrodes 49, 53, 60, 68
Electro-active compounds 21, 28, 52, 61, 87, 88
Electrochemical cell 31
Electrode, capacitance 106
–, conditioning 16, 22, 55
–, preparation 61, 75, 82, 99
–, selectivity 19, 21, 25, 30, 158
–, test 158
Electrometer amplifier 110
Electrostatic induction 117
EMF measurements 108
End-point of titration 128
Enzyme electrodes 99
Equivalent circuit, cell with liquid junction 105
Error, EMF measurement 106
–, general 194
–, inproper input impedance 110
–, insufficient selectivity 128
–, nonlinearity of meter 112
–, standard-addition method 138
–, temperature variation 185
–, titration method 128
Exchange current density 16, 17, 20, 106, 108

Filling solution 69, 175
Flow-thru cells 180
Fluoride determination 166
– electrode 60, 66, 135
Free ion determination 73

Galvani potential 31, 106, 183
Gas sensors 93
Glass membrane electrodes 51
Glucose sensor 101
Glutamine sensor 100
Gold determination 167
Gran plot method 147
Ground loops 119, 182
Grounded solution 107, 182
Guarded signal wire 176

Halide ion electrodes 58, 74
HALTAFALL program, evaluation of titration curves 130
–, gas sensors 97
Handling, glass membrane electrodes 55
–, heterogeneous solid-state membrane electrodes 77
–, homogeneous solid-state membrane electrodes 68
–, liquid and solid matrix membrane electrodes 87
Helmholtz plane 7
Henderson equation 37
Henry's law 95
Heparin, complex formation with Ca^{2+}ions 172
Heterogeneous solid-state membrane electrodes 74
Homogeneous solid-state membrane electrodes 58
Hydrofluoric acid sensor 93

Hydrogen electrode 15, 32
– cyanide sensor 93
Hydrogenphosphate electrode 88
Hydrogen sulfide sensor 93
Hyperparathymoism 172

Indifferent electrolyte 30, 42, 124
Indirect determination 141, 188
Individual activity coefficient 121, 197
Industrial filter 181
Input current 90
– impedance 90
Insecticides determination 99
Insulation problems 116, 176
Interferents, elimination of 30
Intracellular measurements 174
In-vitro measurements 171
In-vivo measurements 174
Iodide determination 167
– electrode 63, 74, 76, 91
Ion buffer 63, 123, 173, 190
– meter 111
Ionic strength adjuster 124
Iron determination 66, 167
Isotherms intersection point 183, 194

Kjeldahl method 98

Lanthan determination 167
Lead determination 167
– electrode 58, 79
Leucin sensor 101
Life time, enzyme electrodes 102
–, glass membrane electrodes 57
–, liquid and solid matrix membrane electrodes 81, 89
–, solid-state membrane electrodes 61, 75
Light sensitivity of halide electrodes 59
Lipophilic ion interference 87
Liquid junction potential 35, 66, 74, 107, 122, 124, 173, 187, 188
Liquid membrane electrode 77, 87
Lithium determination 66, 167
– electrode 79
Logarithmic amplifier 125, 126
Low-noise cable 118

Macrotetrolides 27
Magnesium determination 132, 167
– electrode 77, 84
Manganese determination 167
Manufacturers of ion-selective electrodes 210
Measuring current 17, 110
Membrane processes 17
Mercaptans determination 167
Mercury determination 167
– electrode 65
Metal^{2+} electrode 77
Metallic buffers 63
Methionine sensor 101
Microelectrodes 54, 89, 171, 175
Micro technique 133
Mixed potential 22
Molybdenum determination 167
Monitor, Cyanid 188

Nernst equation 11, 121
Nickel determination 168
Nikolsky equation 29
Nitrate determination 98, 168
– electrode 79
Nitrilotriacetic acid determination 168
Nitrite determination 101, 168
Nitrogen determination 98
– oxide sensor 93
Noise level, EMF measurements 114
Non-aqueous media 68, 77, 83

Offset voltage 125
One-point titration 136
Operational amplifier 178
Oscilloscopic measurements 118
Oxalate determination 168
– electrode 90
Oxidizing substances determination 168

Palladium electrode 88
Penicillin sensor 101
Perchlorate determination 168
– electrode 79, 91
Periodate electrode 88
Permanganate electrode 88
Phenylalanine 92, 101
pH-glass membrane 21, 43
Phosphate determination 168
– electrode 87
Photo EMK 59
Poggendorf's method 109
Poisoning of electrodes 68, 77, 87
Polarization of electrodes 106, 182
Potassium determination 168
– –, intracellular 174
– –, blood serum 145, 172
– electrode 79
Potential determining ions 14
– chemical 12, 16, 197
–, electrochemical 12
–, electrostatic 11
–, equilibrium 12
–, Galvani 11

–, streaming 9
–, surface 9
–, Volta 11
Potentiometric titrations 14, 127
Potentiometry, fundamentals 6
–, direct 14
Precipitation titrations 68, 128
Protein interference 173
Pungor electrodes 74
PVC membrane electrodes 81

Quarternary ammonium ion electrode 79

Reactivation of glass membrane electrodes 57, 97
Reference electrodes 31
––, diaphragm cleaning 45
––, leakage rates 40, 45
––, micro 177
Resistance, microelectrode 176
Response time 26, 73, 81
––, Ag_2S electrode 157
––, bio-sensors 102
––, gas sensors 97
––, microelectrodes after C-neutralization 178
Rest potential 177
Rubidium determination 168
– electrode 79

Salicylate electrode 88
Salt bridge 38, 40, 173
Sample addition 140
Selectivity coefficient 29, 159
Selectrode 89
Selenium determination 168
Shunt electrode 31
Silicone rubber membrane electrodes 74
Silver determination 135, 169
– electrode 58, 74, 76
Silver/silver chloride reference electrode 44
Slope 123
–, control 126
Solid-state membrane electrodes 58, 74
Sodium determination 169
––, blood serum 145
– electrode 52, 79
Solubility, Ag^+ in KCl 44
Standard addition method 137, 142, 187
– electrode potential 33
– hydrogen electrode 32
– subtraction method 138
Sterilization of electrodes 174
Storage of bio-sensors 102
–, gas sensors 97
–, ion-selective electrodes 87
Stray currents 107
Streaming potential 66, 180
Strontium electrode 77, 84
Sulfate determination 101, 169, 188
– electrode 90
Sulfide determination 169
– electrode 58, 74, 76
Sulfite determination 169
Sulfur determination 169
Sulfur dioxide determination 170
––, free and bound in beverages 98
––, sensor 93
Supporting membrane material 88
Surface Potential 106
Symmetric cell 183

Tellurium determination 170
Temperature coefficient, Volta Potential 106
––, electrochemical cell 183
– compensation 125, 183, 194
– hysteresis, reference electrodes 45, 47, 48
Tetraalkylammonium electrode 79
Tetrachloroferrate electrode 88
Tetrachloromercurate electrode 88
Tetrachloropalladate electrode 88
Tetrachlorozincate electrode 88
Tetrafluoroborate electrode 79
Tetraiodomercurate electrode 88
Thalamid reference electrode 46
Thallium determination 170
– electrode 51, 88
Thiocyanate determination 170
– electrode 63, 74, 76, 87, 91
Thiol group determination 170
Thiosulfate determination 170
Thorium determination 66, 170
Tip potential, microelectrodes 177
Titration 127, 191
Total carbon determination 98
Trace analysis 73, 103
Transfer coefficient 19
Transistor electrode 194
Transient response 24
Tungsten determination 170
Tyrosine sensor 101

Unsymmetrical cell 185
Uranium determination 88, 170
Uranyltribenzoate electrode 88
Urea sensor 99

Valinomycin 28, 79
Vanadium determination 170
Viscosity of electro-active phase 88
Volta potential 106

Water hardness 77, 170

Yeast 101

Zero point potentiometry 133
Zinc determination 170
Zirconium determination 170

Index of Symbols Used

A	work, ampere (current)
Å	ångström (10^{-8} cm)
a_B	activity of component B
b	bandwidth of an amplifier in Hz
c	capacitance
C_B	concentration of component B
d	distance
$\mathbf{E}$	EMF, electric field strength (vector)
e	electron
e_0	elementary charge
F	Faraday constant
f	frequency, activity coefficient
G	Gibbs free energy of reaction
$G^{\ddagger}$	Gibbs free activation energy
H_o	outer Helmholtz plane
H_i	inner Helmholtz plane
h	Planck's constant
I	ionic strength
i	current density
i_0	exchange current density
i_m	measuring circuit current
K_S	solubility product
K_{CF}	complex formation constant
k^{pot}_{M-I}	potentiometric selectivity coefficient
k	Boltzmann constant
N_A	Avagodro's number
n	number of moles
p	pressure
q	electric charge
R	gas constant, resistance
S	electrode slope
$\mathbf{s}$	path (vector)
T	absolute temperature
t_i	transport number of ion i
U	electrical voltage
V	volt, volume
x	mole fraction
z	ionic charge
β	constant, transfer coefficient
ϵ	dielectric constant, electrode potential
ξ	electrokinetic potential
κ	Debye-Hückel reciprocal length
μ	chemical potential
$\bar{\mu}$	electrochemical potential
υ	stoichiometry coefficient
φ	electrostatic potential
ϕ	Galvani potential
χ	surface potential
ψ	Volta potential
$\ominus$	electron in the conduction band of a semiconductor
$\oplus$	hole in the valence band of a semiconductor

MIX
Papier aus verantwortungsvollen Quellen
Paper from responsible sources
FSC® C105338

If you have any concerns about our products,
you can contact us on
ProductSafety@springernature.com

In case Publisher is established outside the EU,
the EU authorized representative is:
Springer Nature Customer Service Center GmbH
Europaplatz 3, 69115 Heidelberg, Germany

Printed by Libri Plureos GmbH
in Hamburg, Germany